I0820152

WHAT COULD POSSIBLY GO WRONG?

WHAT COULD POSSIBLY GO WRONG?

The Worst Best Ideas You've Never Heard Of

DUNCAN McKENZIE

Published by Collins, an imprint of HarperCollins Publishers Ltd

First Edition

HarperCollins books may be purchased for educational, business, or sales promotional use through our Special Markets Department.

HarperCollins Publishers Ltd
Bay Adelaide Centre, East Tower
22 Adelaide Street West, 41st Floor
Toronto, Ontario, Canada
M5H 4E3

www.harpercollins.ca

HarperCollins Publishers
Macken House, 39/40 Mayor Street Upper
Dublin 1, D01 C9W8, Ireland
https://www.harpercollins.com

Library and Archives Canada Cataloguing in Publication

Title: What could possibly go wrong?: the worst best ideas you've never heard of / Duncan McKenzie.
Names: McKenzie, Duncan (Comedy writer), author.
Description: First edition. | Includes bibliographical references.
Identifiers: Canadiana (print) 20250234475 | Canadiana (ebook) 20250238152 | ISBN 9781443475167 (softcover) | ISBN 9781443475174 (ebook)
Subjects: LCSH: Inventions—Humor. | LCSH: Inventions—Miscellanea. | LCGFT: Humor. | LCGFT: Trivia and miscellanea.

Classification: LCC T212 .M35 2025 | DDC 602/.07—dc23

ISBN 978-1-4434-7516-7

Printed and bound in the United States of America

25 26 27 28 29 LBC 5 4 3 2 1

CONTENTS

WHAT COULD POSSIBLY GO WRONG?

INTRODUCTION

SOMEONE has come up with new idea or invention! What could possibly go wrong?

Sometimes an idea is dangerous, like the flying suit that would allow its inventor to jump from the Eiffel Tower, spread his wings, and land gently on the ground below. They say "two out of three ain't bad"—but it was in this case.

At other times, the idea is great but getting it to market is an uphill battle. A woman who tried to roll out an amazing sapphire needle for record players didn't realize how low her competitors would stoop to block her.

Some ideas, like the combination "automatic cat flap and nuclear launch system," were bonkers to start with. Others were good ideas but not quite thought through—like Edison's horrific talking doll.

The world of inventions and creative new ideas is also a world of desperation, jealousy, suspicion, and madness.

So, what could possibly go wrong? As you'll see, the answer is: almost everything.

CHAPTER 1

A BETTER-ER MOUSETRAP

"**BUILD** a better mousetrap, and the world will beat a path to your door."

That saying became popular in the 1800s. The idea is simple enough: if you can come up with something new and useful, even something as humble as a mousetrap, it will attract customers, and you will do well from it.

The saying came from the brain of writer and lecturer Ralph Waldo Emerson. The original version, from 1855, went like this: "If a man has good corn or wood, or boards, or pigs, to sell, or can make better chairs or knives, crucibles or church organs, than anybody else, you will find a broad hard-beaten road to his house, though it be in the woods."

What's missing from this quote? Mousetraps, that's what. Emerson didn't mention them at all. What's more, he wasn't talking about how to make a profitable invention, but about the way that a person can build a reputation and win respect and fame just by doing one thing really well.

The version of the famous quote using the words "a better mousetrap" first appeared in print in 1882, after Emerson died. It was probably a reporter's misquote, but it stuck, and in the years since, people have been taking it way too literally—especially in the United States. Since the mid-1800s, hopeful American inventors have filed more than 4,000 patents for mousetrap designs alone, making it the single most patented item. Every year, the US patent office grants around forty new mousetrap patents.

As we'll see, most of these "better mousetraps" didn't cause crowds of admirers to beat a pathway to the inventor's door.

EARLY MOUSETRAPS

In 1879, James M. Keep of New York patented the "Royal No. 1" mousetrap. It was a sturdy metal device designed to be used over and over. When a mouse approached the trap and poked at the food holder, the jaws snapped shut, gripping the rodent between two sets of long metal teeth. The cast-iron trap had an ornamental design: an incongruous heart motif crushed the poor creature's head.

Some people collect mousetraps, and a Royal No. 1 is one of the most desirable. An example in good condition is worth thousands.

People sometimes say the Royal No. 1 is "the first mousetrap," but that's not close to being true—it's only the first one patented in America.

A few years earlier, in 1860, British inventor Colin Pullinger designed a mousetrap that was much more effective than most of the snap traps that followed. It didn't need resetting—it could catch mouse after mouse. It used a teeter-totter mechanism: when the mouse entered to investigate the smell of the bait, a rocker tipped behind it, blocking the exit. The mouse could only continue through a one-way wire gate, and it would then find itself trapped in a chamber at the end of the device. Once the mouse had passed over the teeter-totter, the barrier rocked back to its original position, ready for the next mouse. It was up to the owner to figure out what to do with the captured live mice.

The mousetraps were very popular, and Pullinger soon had a factory employing forty people to build them. Pullinger said that he had used his "perpetual mousetrap" to catch twenty-eight mice in a single night, and that others had caught over a thousand mice in a month.

A few years ago, one of these antique traps was on display at a museum. One day, workers noticed that it held a dead mouse that hadn't been there a few weeks earlier. Pullinger wasn't kidding when he called his trap "perpetual."

Another type of mechanical mousetrap looked like a paddlewheel with four sections. Each one was smeared with bait. When the mouse stepped onto a paddle, the wheel turned, dropping the mouse into a bucket below and moving the next paddle into position. Again, one trap could catch many mice. This design goes back at least 500 years.

Snap traps aren't new either. In the Middle Ages, this type of trap was often made of wood and killed the mouse by slamming a wooden block down into a baited frame below. Instead of a metal spring, it used a loop of cord with a stick in the middle, twisted so tightly that it exerted a strong pull. Other versions of this trap didn't bother with a twisted cord—they just put a heavy rock on the board.

The fact is, it's not that difficult to trap a mouse. People have been inventing mousetraps for thousands of years, usually by modifying the traps they used for catching birds and small animals. Mousetraps have been found from ancient Egypt—their designs also used a twisted cord for a spring. When the mouse touched the bait, two curved wooden jaws snapped together. Modern reconstructions work very well—and so, apparently, did the originals. In 2020, a team of archaeologists reported finding a tiny package in an Egyptian tomb. It was a mummified shrew. When they X-rayed its bones, they deduced that it had been killed by one of these ancient mousetraps.

Build a better mousetrap, and the world will beat a path to your pyramid.

PROBLEM SOLVED

In 1894, twelve years after "build a better mousetrap" first appeared in print, William C. Hooker of Abingdon, Illinois, invented the definitive "better mousetrap." It's the iconic mousetrap we still use today, and his original version is almost identical to modern designs—the same rectangular wooden base, the same spring and wire "kill bar." Hooker's mousetrap wasn't fancy.

The speed and power of the spring mechanism makes it very effective in killing mice. It was designed from the start to respond with deadly force to the lightest touch.

The small size of Hooker's mousetrap was one of its biggest selling points. The design was more discreet than most mousetraps, and homeowners could tuck it away in a corner or slide it under the legs of a cabinet. Hooker named his trap for this feature—he called it the "Out O' Sight" mousetrap.

Another big advantage to Hooker's mousetrap was its low cost. The trap was easy to manufacture from very cheap materials. For the price of a single all-metal trap, consumers could buy many of these wood-based traps. They became even more affordable over the years, as the traps were produced in larger numbers. The traps eventually became so cheap that people who caught a mouse didn't bother removing the mouse and resetting the trap. Instead, they would just throw trap and mouse straight into the garbage. That was good for mousetrap sales.

Other inventors copied Hooker's design—an American inventor named John Mast sometimes gets the credit for thinking up this type of mousetrap, but he only made tiny changes to Hooker's design, then patented his version independently.

A British inventor named Henry Atkinson came up with another variation, sold under the name "Little Nipper." But they were all basically the same as Hooker's original.

Hooker knew that others were imitating his invention, but he didn't seem too bothered about it. He may have figured that, after coming up with that first successful mousetrap, he could do even better with a second. Sure enough, his next mousetrap simplified the design even further. He did away with the fussy metal trigger that holds the bait. Instead, he moved the triggering mechanism to the underside of the trap. The wooden base was designed to rock slightly. You could put your bait anywhere on the front of the trap. If the mouse stepped on the wood, the trap tipped forward, releasing the kill bar.

It was a nice improvement and made for a wonderfully simple trap—a true "better mousetrap." But nobody seemed to care—the previous design had worked just fine, and they were used to it. The public ignored the "Mk. 2" and have been using the original version—invented by Hooker and copied by many others—for the past century and a half.

A BETTER-ER MOUSETRAP

After people had invented and mass-produced what is pretty much the ideal mousetrap, you'd think inventors would stop trying to come up with a better one.

Not at all.

That "better mousetrap" quote was even better as a mindtrap. Over 4,000 mousetrap designs have been patented in the United States. The vast majority are from novice inventors, and for every mousetrap patent, it's estimated that there are ten applications that have been refused. That's a lot of failed mousetrap inventors.

The US patent office has thirty-nine official categories of mousetrap, including "smiting," "swinging striker," "impaling," "nonreturn entrance," "choking or squeezing," "constricting noose," and the alarming "electrocuting and explosive." Only around twenty mousetrap designs have been profitable in the stores, so there are more *categories* of mousetrap than commercially successful designs.

A problem with some mousetraps is that they seem more sadistic than effective, using descending spikes, saws, and blades. We won't go into the gory details.

Another common problem is mousetraps that are overly complex—sometimes ridiculously so.

In 1873, an inventor from Pennsylvania, John Gould, designed a very convoluted model. It used a complex set of clockwork gears, rods, and levers to send the mouse into a storage chamber then reset the trap. There were already much simpler traps that accomplished the same thing, so this was like using a bulldozer to crack a nut.

In 1915, John Gassaway from Oklahoma came up with a different clockwork trap—sized for rats rather than mice. It snapped down on the rodent, then flipped the body away as it reset itself. The mechanism might have kept the trap operating, but the user would face a pile of dead rats that the machine had tossed aside.

But not every complex trap was a failure. One inventor who did manage to profit from an overcomplicated mousetrap system was Marvin Glass of Chicago. His design, first sold in 1963 and patented in 1967, used a complex sequence of weights, ratchets, and levers to catch mice under a descending cage. However, he was just making fun of over-complex designs. His mechanism, inspired by contraptions from American cartoonist Rube Goldberg, was created for the wildly successful board game Mouse Trap.

HUMANE TRAPS

Many traps are humane, catching mice alive, so the little darlings can be released into the wild. This doesn't always work so well in practice—mice are usually caught in the middle of the night and often make noisy exertions trying to free themselves. Disposing of the mice legally can also be a problem—it's not illegal to catch live rodents, but many municipalities prohibit releasing them, and if they're not placed far away from the home, they will try to return.

One trap, dating from the 1860s, takes the humane trap to an extreme, turning pest into pet. When the mouse enters the trap, a spring door closes behind it. It is then trapped in a hamster-style cage, complete with food and an exercise wheel. The trap was sold as a novelty item.

Another variant, patented in 1903, had wheels on each side. The mouse could entertain itself by running in its exercise wheel, and its efforts would roll the trap around the room. Again, this was (we hope) more of a novelty item than a serious attempt to control an invasion of wild mice.

BELLING THE RAT

An old folk tale tells the story of a group of worried mice who meet to discuss a problem. They are terrified of a certain cat, which has been killing members of their community. One mouse has a great idea—they will tie a bell around the cat's neck, so they can always hear the cat coming. The other mice applaud the plan, until someone asks the question, "Which of us will put the bell on the cat?"

Perhaps it was this story that inspired an inventing team, American Joseph Barad and his Russian partner Edward E. Markoff, in their 1908 patent for a mousetrap. The device is a cage with a hole small enough for a mouse to put its head through. When the mouse touches the bait on the far side, the trap releases a coiled metal spring, which snaps around the animal's neck. It doesn't hurt the mouse, but the collar has a bell attached, so the mouse will make a jingling sound wherever it goes.

So far, it may not sound like the trap accomplishes much. The inventors freely admit that their trap doesn't catch animals alive, nor does it kill them. In fact, it is designed to release the mouse and let it run free. But once the trap is used, it will remove all mice in the area.

The inventors explain a "well-known fact" that mice and other rodents are "naturally excessively sly and distrustful, even to members of their own family . . . It is also known that the sound or tinkling of a bell is as a rule very terrifying to animals of the species named and that if pursued by such sounds, they will immediately vacate their haunts and homes, never to return."

You can probably see where the inventors were going with their "well-known facts." The mouse with the bell is supposed to return to its nest, where it will meet the rest of its sly, distrusting family. When the other mice hear the tinkling of the bell, they are filled with terror and immediately pack up and leave the area forever. The trap doesn't just remove one mouse—it removes them all.

The invention was designed to work against both rats and mice. In reality, it would be difficult to put a collar around a mouse's head without the mouse wriggling out of it, but it certainly fits well in the diagrams accompanying the patent. Their terrified, running mouse looks like a cross between a pig and a greyhound.

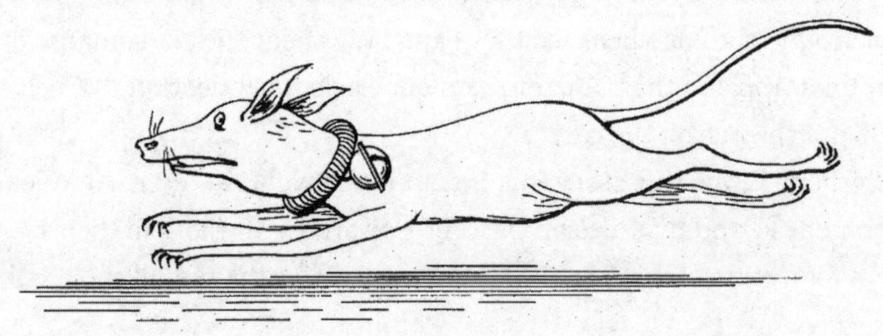

TEXAS MOUSETRAP

One of the more extreme traps was patented by James Williams of Texas in 1882. It was designed for use against mice and other burrowing animals. An illustration in his patent shows it being used against a rat.

Williams wanted to design a device that alerted the owner when an animal had been killed, so the trap could be reset.

When the animal approaches the bait, it walks onto a plate. The plate descends beneath the weight of the creature, tripping a spring-loaded rod. The rod flies back and kills the animal.

Oh, we forgot to mention that the rod is attached to the trigger of a large, cocked, military revolver (to be provided by the user). The pistol is fixed to the trap, pointing down at the place where the little animal will appear. So, watch your fingers when you place that cheese.

Williams was quick to point out the versatility of his trap. He explains, "This invention may also be used in connection with a door or window, so as to kill any person or thing opening the door or window to which it is attached."

With a mousetrap like that, people better not beat a path to *his* door!

CHAPTER 2

THE CRIME FIGHTERS

THE shout of "Eureka!" from a bathtub is one of the most famous moments in inventing history. It signalled the moment when Archimedes figured out how to prove the guilt of a fraudulent goldsmith.

Archimedes had gone from inventor to crime fighter, and he inspired generations of others. Many later inventors have seen themselves as intellectual superheroes, devising ways to deter robbers and thieves, or even bring them to justice.

For some, it was about the challenge of pitting their ingenuity against unseen criminal adversaries. For others, the problem was simpler—while they were gazing into the sky dreaming up new ideas, people kept stealing their stuff.

A POINTY HAT

In 1914, Frank L. Snow of Los Angeles was no stranger to the heartbreak of hat theft. On several occasions, he had left his expensive hat on a hat rack when entering a restaurant, but later, when the meal was over, he'd discover the hat was gone—another diner had walked off with it. He was tired of being a victim—and determined to settle the score with the thieves.

Like an avenging, inventing angel, Snow crafted an anti-hat-theft device that would protect his own hat, and those of other law-abiding citizens, while bringing his wrath down upon the heads of the hat thieves.

It consisted of a small metal disk mounted discreetly inside the rim of the hat. Before the honest hat owner hung up their derby, they used a small key to turn the disk, changing its status from "safe" to "armed." A pointed blade rotated and locked into place. The presence of the device wasn't obvious, but any criminal trying to take and wear the hat would be instantly and painfully stabbed in the head by the metal spike, and their cries of pain would draw more attention than any alarm. Another criminal caught red-hatted!

Naturally, the legitimate owner knew about the mechanism, and when they wanted to put their hat back on, they could flip the device back into the "safe" position in moments. As holder of the key, they alone could wear the hat with no risk of injury.

That is, unless they forgot—in which case their scalp, too, would be stabbed by the cruel blade.

And, of course, they might also be injured when turning the sharp blade to set up or deactivate the mechanism.

But aside from that, it was foolproof.

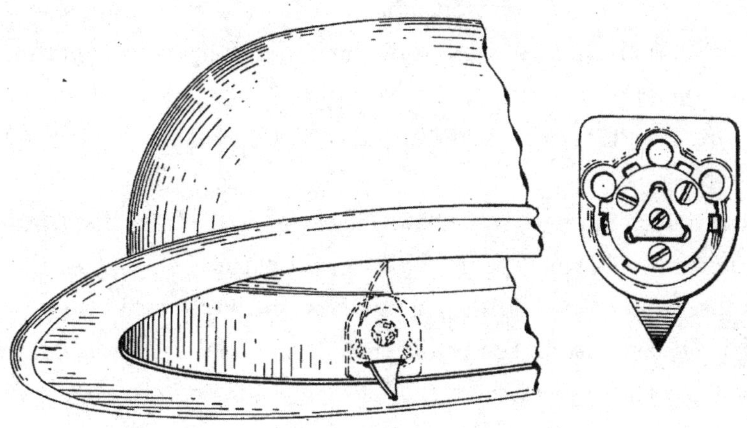

CREAMINAL ACTIVITY

Inventors from Britain showed a special interest in anti-crime devices. In 1955, Frederick Henderson of Dalton-in-Furness, Lancashire, focused on petty crimes rather than great ones. Perhaps he, too, had been a victim of crime.

In Henderson's case, the crime was milk theft.

Refrigerators were uncommon in 1950s Britain, and pint bottles of milk were delivered to most homes every morning. Milk was usually left on the front doorstep. Nothing prevented someone coming along after the milkman had done his rounds and stealing a bottle or two.

Henderson's idea was to fasten a cylindrical wire cage to the top of the front door. The milk bottles would be stored inside it, protected from thieves.

Using the cage involved some changes in the milkman's routine. Instead of placing the bottles carelessly on the doorstep, he would need to insert them into the underside of the cage. A set of one-way flaps allowed bottles to be pushed in from below, one after the other.

The top of the door frame prevented the bottles from being lifted from the top of the cage, confounding milk thieves. When the homeowner unlocked and opened the front door to collect the milk, the top of the cage would no longer be covered by the upper door frame, and the milk bottles could be freely lifted out from above

Using Henderson's invention, milk theft would surely be a thing of the past.

The design did have some flaws, however. A thief with just a little determination could easily push on a milk bottle from below, press the cage flaps open, and lower a bottle from the bottom of the stack.

The system also posed challenges for the homeowner: to lift a milk bottle from the cage at the top of a door, you'd need to be as tall as a basketball player.

Then again, perhaps the occupants could reach that height—if they drank their milk.

A CASE OF DECEPTION

Other British inventors turned their attention to fighting more serious crimes. During the 1950s and 1960s, Britain had experienced a surge in the number of daring robberies. But the scientists and "boffins" who had designed gadgets to help fight Germany now turned their attention to fighting domestic crime.

An inventor named John Fisher came up with a briefcase to protect cash and valuables from theft. It was manufactured and sold under the brand name "Arrestor."

The case looked like an ordinary suitcase, but if a thief tried to snatch it, three sturdy, telescoping rods extended forcefully from the body of the case, forming a 3.5-metre (12-foot) Y shape. With the rods extended, it was difficult to escape with the case and impossible to fit it through a doorway or inside a car.

At this point, any sensible crook would likely drop his prize and escape. Case closed . . . and dropped on the ground. But Fisher believed in justice—the thief must not be allowed to get away. In the same instant that the rods extended, the handle of the case snapped powerfully back into the body of the case, trapping (and likely crushing) the thief's fingers. The criminal who had grabbed the case would find himself unable to release it.

In addition to deploying these pneumatic devices, the case also sounded an alarm—hopefully one loud enough to be heard over the thief's screams of pain.

Once police had arrived on the scene and had the thief in hand, the rightful custodian of the case could use a special key to reset the case: the alarm was deactivated, the rods were retracted, and the criminal's fingers were removed—one way or another.

Newsreel footage of the invention still exists. In a demonstration from 1961, it performed well—the rods blast out with impressive power—but the sturdy mechanism took up a good deal of the useful space inside the case, limiting its capacity.

Curiously, the inventor didn't do much to stop the case from being stolen in the first place—for example, by using a chain around the carrier's wrist. We have a sneaking suspicion that he was more interested in outsmarting and trapping unsuspecting thieves than in actually protecting the valuables.

Fisher invented an accessory to go with his briefcase: it was a black bowler hat, externally identical to the type worn by most British businessmen of the era, but lined with steel. The 1961 newsreel shows a businessman calmly reading a newspaper, hardly seeming to notice as attackers strike the top of his hat repeatedly with a club—although a real robber might think to take a shot at the unprotected side of his head instead.

Fisher wasn't the first person to think of a steel-reinforced hat. It's said that supervisors in Belfast shipyards also reinforced their bowler hats with steel to protect against resentful underlings who might "accidentally" drop a tool on the boss's head.

Fisher's steel hat and gadget-loaded case were not commercial successes, but they seem to have inspired a generation of screenwriters: two years after Fisher's briefcase was unveiled, a booby-trapped briefcase featured in the 1963 James Bond film *From Russia with Love,* while in *Goldfinger* a steel-lined bowler hat became a deadly weapon for Goldfinger's henchman Oddjob. A steel bowler was also standard equipment for dapper agent John Steed, in the 1960s TV series *The Avengers*.

MURDER IN THEIR EYES

Some inventors have extended their crime-fighting ambitions to murder. Even today, with modern technology, it is often difficult to identify a killer. The problem was much worse back in the 1800s, a time when many murders went unsolved. The quest to improve detection led to a strange and flawed technology.

In those years, many people believed that when a person died, the last thing they saw would be fixed onto the retina. It was even said that if you gazed into the dead eyes of a murdered person, you could see the face of their killer (which could be true, but only if you were the murderer, looking at your reflection).

In the mid-1800s, a British photographer named William H. Warner carried out experiments to examine this odd belief. He went to a slaughterhouse and took the eye of a slaughtered calf, then photographed the retina. When he looked at the photographs under a microscope, he could see lines. These, he was sure, were the lines on the pavement of the slaughterhouse floor. He believed the image he saw on the retina was the calf's final glimpse of the world. He was sure his discovery could be valuable in solving murders.

In April 1863, Warner had a chance to put his theory to the test. A woman named Emma Jackson had been murdered in the London neighbourhood of St. Giles, her throat cut by her killer. Jackson had worked as a shirtmaker and occasionally supplemented her low income through sex work. It appeared that she had been killed by one of her male clients. The public was appalled by the violence of the murder. A detective named James F. Thomson was assigned to the case.

Warner got in touch with Thompson, reminding him of the well-known fact about the retina, and described his experiment with the calf's eyes. He offered his services in examining the eye of Emma Jackson—which might reveal the face of her killer.

Thomson wrote back thanking Warner, but explaining that, on this occasion, photography wouldn't help. Like Warner, Thomson had also thought about solving crimes by examining eyeballs. He told Warner that, a few years earlier, he had talked to an eye expert about the method and was told that any images on the retina would only remain for twenty-four hours before they faded away completely. Unfortunately, Emma Jackson had died at least forty hours ago and had since been buried.

So, despite police enthusiasm, and the support of a professional photographer, retinal photographs didn't help crack the case. In fact, nothing did. The murder was never solved.

Still, plenty of people believed that examining eyeballs might someday be a way of solving murders. In 1887, a scientific breakthrough in Germany seemed to offer new hope. A professor named Franz Boll studied a chemical called rhodopsin, which is found in the retina. Rhodopsin changes colour from purple to transparent when it is exposed to light. Boll drew the reasonable (though wrong) conclusion that this light-sensitive substance was how we see—the eye must work like the photographic plate on a camera.

One of Boll's colleagues at the University of Heidelberg, Willy Kühne, took the work further. He believed rhodopsin was the key to capturing images from human eyes. Kühne carried out experiments on frogs and then rabbits. He kept the rabbits in the dark, causing the rhodopsin in their eyes to change slowly from transparent to purple, then he held them in one place facing a window. After the rabbit had been exposed to the window's light for a few minutes, he quickly decapitated the unfortunate creature so he could examine its eyes. After a little chemical tinkering, he produced photographs of the rabbit retina, which clearly showed images of his lab window as seen through the eye of a rabbit.

The public were amazed by Kühne's results. His research appeared to provide scientific support for an idea that many people already believed—that eyes could see because they had pictures inside them. Kühne's work seemed to suggest that the pictures were continually reset in a living animal, but that the final dying vision remained for a time after death and could be fixed and preserved. Once again, it looked as if the world was on the verge of being able to capture the last thing a murder victim saw. Kühne was pioneering an exciting new science, which he called "optography."

Throwing ethical considerations to the wind, Kühne looked for ways to try out his methods on a human subject. In 1880, a German murderer had been sentenced to execution by guillotine. Kühne received official permission to examine the man's eyes after death. He waited eagerly for the head to drop, and once he had it, he scurried back to his lab to remove the dead man's eyes. The results weren't too impressive—the image seemed to show a vaguely rectangular shape—some saw it as the outline of a guillotine blade, but since the murderer was blindfolded when he was killed, it's more likely the image was nothing at all.

In hindsight, we know that Kühne and his colleagues didn't understand what they were seeing. The main function of the purple-red chemical they had discovered is to make the eye more sensitive in dark conditions. It takes about thirty minutes to appear, and when it does, it makes the retina much more sensitive to light—allowing people (and other animals) to see in the dark. If the lights are turned on, rhodopsin becomes transparent in a fraction of a second, and you lose your night vision until the chemical has regenerated.

Kühne's methods of optography only worked under unusual conditions. He needed to keep his rabbits in the dark, then hold them absolutely still as they were uncovered and pointed at the window. Kühne had unwittingly used rhodopsin to produce something like a biological camera—the tantalizing results were just an artifact of his experiment, and they had very little to do with how eyes worked under normal circumstances. Eyes don't function like cameras, and an experimenter couldn't normally see anything on the retina that would show what an animal had been looking at.

Still, the scientific-looking methods created excitement. The inventor Nikola Tesla was no stranger to odd ideas. He latched onto this one and wanted to take it further. Tesla figured that, since images land on the retina and end up as images in the mind, the same must work in the opposite direction—ideas in the mind must create pictures on the retina. He announced his own invention—a machine that would magnify the retina and project it onto a screen, so everyone could see your thoughts.

Some Tesla fans have seen this as an idea years ahead of its time. But its time is "never"—humans don't show their thoughts on the backs of their eyeballs. If they did, optometrists would have a heyday checking your retinas and looking into your mind.

It took decades for people to realize that dying visions are not preserved inside the eyeballs. Sensational newspaper articles continued to describe new "scientific discoveries" based on this falsehood. Some murderers believed the technology worked and seem to have deliberately destroyed the eyes of their victims out of fear that the retinas could be used as incriminating evidence. Fortunately, courts remained wary about evidence from the retina, and optograms were never successfully used to convict anyone.

SAVE THE "CAT"

Modern inventions to prevent crime are often high-tech, like drones and body scanners. But sometimes an innovative solution can be much simpler.

One expensive modern crime is catalytic converter theft. A catalytic converter is found on the exhaust system of most gasoline-powered cars. The device is basically a small oven that cooks the gases emitted by an engine. Very toxic gases go in, and less toxic ones come out.

The device carries out its chemical baking using a catalyst—a substance that works like a chemical knife, continually cutting up the molecules in the exhaust. If you use a knife to cut up ingredients for a pie, you still have the knife when you're done, and a catalyst works the same way—it's an implement rather than an ingredient. And like a quality knife, a catalyst in a car can be expensive—it's usually made from a precious metal like platinum.

The quantity of precious metal inside the catalytic converter is tiny, but it's still enough to make the part a tempting target for thieves. A single catalytic converter can be sold to an unscrupulous scrap metal dealer for a couple of hundred dollars. The loss to the car's owner is much higher—replacing the part costs thousands.

Two Canadian women, Mavis Shaw and her daughter Tamara Dolinsky, had the catalytic converter stolen from their vehicle. They were warned by the insurance company that it was likely to happen again, so they thought about ways to protect their car.

Many inventors have thought about ways to tackle this common crime. Deterrence methods include setting up camera systems, placing a cage around the converter, installing a shield over it, etching the car's VIN onto its metal, or adding a car alarm just for that one part.

But Shaw and Dolinsky found a novel approach. They designed a protective metal fence that can be quickly assembled around a car, no matter where it's parked. The fence is only a few inches high and painted bright yellow. The car's tires rest on metal plates welded to the fence. Once the car is driven onto the plates, the weight of the vehicle holds the barrier firmly in place.

It doesn't make theft impossible—nothing can do that—but it certainly makes it difficult for a thief to slide under the car and access the catalytic converter. Once the inventors' own vehicle was repaired, they placed the fence around it, and according to news reports, it has remained unmolested.

The women heard about a "catalytic converter challenge," run by the Edmonton police and an insurance company. They submitted their invention to the contest. The judges were impressed—the invention was simple, original, and clever. It seemed like a great deterrent.

Will car owners everywhere install this fence around their cars every time they park? Perhaps not, but it might be worthwhile for those who must park in high-risk areas. And regardless of whether their invention takes off in the marketplace, it quickly produced financial benefits for Shaw and Dolinsky—not only were they spared further car repairs, but their first-place win in the competition earned them a $25,000 prize.

CHAPTER 3

STAY SAFE

WHEN inventors aren't making the world more dangerous for mice or criminals, they often think about making it safer for humans.

Some of these innovations have made a difference. Car seatbelts and airbags have unquestionably reduced deaths on the roads, and an emphasis on safety in the airline industry has turned air travel from a very dangerous mode of transformation to a very safe one.

But not every life-saving idea actually saves lives. In fact, some have proved deadly.

DON'T JUMP JUST YET

Benjamin Oppenheimer felt a sense of terror when he looked at the world around him. It was 1879, and more and more people were living in big cities, with the latest skyscrapers soaring up to ten floors above the ground. People could now reach the top floor using newfangled, steam-powered elevators. But what if there should be a fire? How would people who lived on the top floors manage to escape from the burning building?

Oppenheimer's solution to this nightmare scenario was a parachute helmet. The stranded apartment dweller put the helmet on and tied the strap firmly under their chin. Four strong cords were attached to the helmet connecting it to a parachute.

Two padded boots (a "paraboots"?) were supposed to help soften the impact of the landing.

After donning this gear, the high-rise dweller only had to step out of the apartment window and drift slowly to the ground.

The invention was flawed, to say the least.

One obvious hazard was the way the parachute attached—it was held on only by the helmet's chin strap, so the jumper was likely to be choked as they fell.

But a bigger concern is the parachute itself. When people jump from a plane, their parachutes are usually around 9 metres in diameter, but this one was a piece of cloth fastened around a permanent rigid frame "about 4 or 5 feet in diameter." That's a small parachute.

It's unlikely Oppenheimer's parachute was ever used, but if it had been, would it have worked at all? It's an interesting problem. In Benjamin Oppenheimer's day, the tallest skyscraper was the ten-floor New York Tribune Building, which was nearly 80 metres tall. What would happen if a person jumped from the top, wearing this parachute hat?

To find out, we contacted engineer and inventor Alan Eliasen, whose programming language Frink was designed for exactly such problems as this. According to Eliasen's calculations, if the parachute stayed in one piece, it would slow the person's fall considerably. When the parachutist reaches the ground, five seconds later, the impact of the 80-metre fall will have been reduced to that of a mere 27-metre drop. That's a surprisingly good reduction. The bad news is that, according to most experts, any fall from more than 24 metres is almost certainly fatal. In this case, the jumper will hit the ground at more than 80 kilometres per hour. Wearing the inventor's padded shoes will only mean they land with a slightly softer SPLAT.

A SUIT FOR THE FALL

In the early 1900s, a safety-conscious inventor in France fretted about the well-being of the daredevils who flew in aircraft.

After the Wright brothers made their first successful flight in 1903, aviation had literally taken off. People were flying further every year, including a flight across the English Channel in 1909. But if a technical problem developed on a longer flight, the pilot still needed to land the plane or die trying. There was no means of safely abandoning the aircraft in flight.

Franz Reichelt was a Bohemian who had settled in France, and he became interested in this aspect of air safety. Other inventors were also trying to develop parachutes, but their creations were still too big and bulky, and they only worked if you were jumping from a high altitude.

Reichelt envisioned a kind of winged suit. A pilot could wear it in the plane—its folds of loose fabric would flop around them but wouldn't inhibit their movements. If the pilot needed to leave the plane suddenly, they could just leap out—the suit would instantly catch the air and slow their fall. Even at a low altitude, the pilot would float safely to the ground.

Reichelt was a tailor by profession, so he stitched together a suit. He sent it to a French aviation group for testing, but they weren't interested—they didn't believe his design would work.

Their rejection irked Reichelt. He was sure his suit would work, and his enthusiasm to prove it grew stronger when an aviation enthusiast offered a prize of 10,000 francs (equivalent to five years' earnings for a typical French tailor) to anyone who could develop a working safety parachute for aviators. The competition required that the parachute must weigh no more than 25 kilograms. Reichelt's original design was nearly three times that weight, but he found ways to reduce it while further increasing the surface area.

He tested the para-suit on human-sized dummies, which he dropped from tall buildings. Unfortunately, the mannequins just smashed to the ground. The problem was obvious to Reichelt—they weren't being dropped from high enough. The suit needed more altitude to catch the air, and a typical Paris building wasn't high enough to simulate escape from a plane.

To test his suit properly, Reichelt needed a much taller structure. He set his sights on what was then the tallest building in the world—the Eiffel Tower. He wrote to the authorities, requesting permission to use the tower for his tests. They said no at first, but Reichelt was very persistent, and they finally agreed.

On the morning of February 4, 1912, he drove to the Eiffel Tower to carry out the demonstration, which would be conducted from the first level of the tower, a height of 57 metres—a little under one-fifth of the tower's total height, but still a long way to fall.

Crowds and photographers had gathered to witness the spectacle. Reichelt wore the baggy suit and modelled it for the press cameras and film crews, demonstrating its winglike folds.

What came as a surprise to everyone, however, was that Reichelt had not brought a dummy for the tests. He intended to perform the jump himself. His friends were astonished to hear of his plans and tried to talk him out of it, but he brushed their objections aside.

The police were nervous, too. They didn't want to allow the jump, but it looked as if Reichelt had all the correct paperwork for official approval. (Later, the people who'd signed off on the test said they had assumed it would involve a dummy and would never have given permission for a live jump.)

Another parachute experimenter, Gaston Hervieu, was present for the test. He had carried out multiple experiments dropping mannequins from aircraft. When he learned of Reichelt's intentions, he begged him to reconsider—Hervieu didn't think the tower's first platform was high enough for a parachute to work.

Reichelt insisted Hervieu was wrong: his winged suit would open instantly. He told his rival to stand back and watch because his body and his parachute would provide "the most decisive of denials."

Reichelt went up to the first platform of the tower. A film crew accompanied him, and he stood on the edge of the railing. His suit unfolded into a large semi-circular cloak with a squared-off hood.

A second camera crew waited on the ground, ready to capture the descent.

He stood for some time on the edge of the railing, staring down at the long drop below him and summoning up his courage. It's one thing to believe in your invention, but another to trust your life to it.

Finally, Reichelt made his jump. He dropped like a stone and was killed instantly when he hit the ground.

In a way, he proved his point to Gaston Hervieu—his parachute had opened. The problem was that it didn't work.

A SAFER SLICER

Many safety-related inventions address daily tasks. For example, shaving used to be a dangerous business. There's a reason a straight razor was also called a "cut-throat razor." Scraping an ultra-sharp blade against your neck required care and focus—a mistake could be painful or even deadly. Even a smaller cut could lead to a deadly infection. That's why many men were shaved by a barber years ago, rather than doing the job themselves.

Many inventors looked for ways to create a "safety razor." A common approach was to add guards near the blade of a traditional straight razor. This could reduce the chance of a serious injury, but small nicks were still possible, and sharpening the blade was difficult.

Other safety razors used replaceable blades. Comb-like guards kept skin and fingers away from the sharp edge. The blades were usually sharp on only one edge. This style is rarely used in razors today, but the blade design is still used in scrapers for removing paint and stickers.

Some attempts at safety razors were very odd. In 1900, Samuel Bligh of Pennsylvania patented a device looking like a paint roller but wrapped in sandpaper. Once it was attached to an electric motor, the fast-spinning roller would scrape the pesky hairs away, leaving smooth skin. Or, if you left it on too long, it would scrape the pesky skin away, leaving smooth bone.

Charles Bailey of Los Angeles invented a very different kind of safety razor in 1924. His "electric safety razor" had a battery-powered motor, which turned a metal cylinder. The cylinder had slits along its surface that functioned as sharp blades, scraping directly against the skin. The device had no mesh screen and no guards. So why was it called an electric safety razor? Perhaps because it sounds better than an "electric skin-flaying razor."

Frank Pollifrone came up with a different electric razor in 1928, although perhaps the term "electric razor" is misleading. It was a battery-powered flashlight with a razor head attached to the top. It allowed you to see your skin better and get a good look at each nick when you're shaving in the dark.

In 1932, an inventor named Burton Christmas patented a strange, disposable safety razor. It was designed to look like a standard book of matches and could be printed with a promotional message and given away. Unfolding the flap revealed a razor blade stapled into the area where the matches would normally be fastened. The user was supposed to fold back the blade, then carefully fold in the sides to form handles from the waxed cardboard. If you completed this dangerous origami without cutting your fingers, you would hold a cardboard "safety razor." While it probably didn't provide much safety in shaving, it might be useful if you needed to protect yourself with a close-combat weapon.

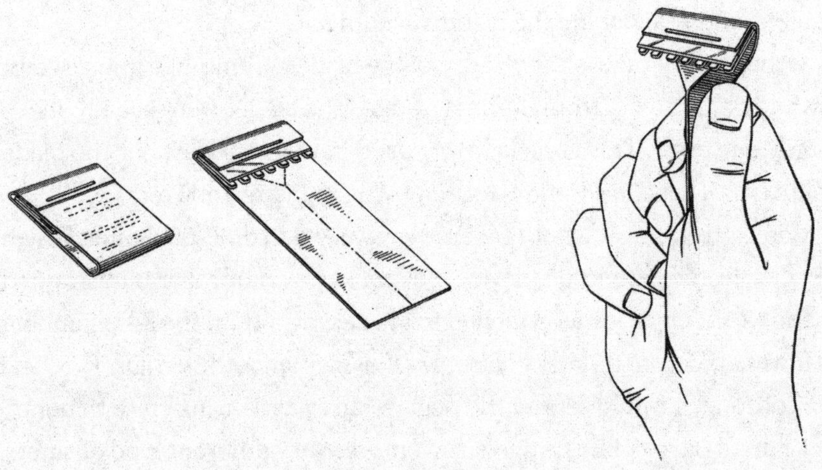

The most successful breakthrough in safety razors came from King Camp Gillette, founder of the Gillette company. He invented a now-familiar razor design with a cylindrical handle and an inexpensive, disposable drop-in blade. As we'll see later, his design took off when it was issued as part of a standard soldier's shaving kit in World War I. The young men might be shelled or shot, but they could march off to war confident that they could shave in safety.

MINE, ALL MINE

Many important inventions have been ways to improve safety in coal mines. Lighting was a particular problem. If you're going into an area full of explosive gas, you probably don't want to light your way with a candle. But that's exactly what coal miners did in the 1800s. Not only could coal dust in the air catch fire, but cracking away at a wall of coal could release pockets of methane, and the result was often an explosion and many dead miners.

The solution miners came up with was as dangerous as it was spectacular—a "fireman," who was often an old miner, would cover himself in layers of wet rags, then crawl into a gas-filled area to ignite it with a candle. It might have created a controlled explosion that burned off the gas, but the wet clothes did not offer good protection, and the method risked killing the fireman.

A famous inventor and scientist named Humphry Davy thought he could come up with a better solution. In 1815, he looked at ways to stop the flames from miners' lights from igniting the air around them. He carried out careful experiments to test how the explosions started. He discovered that if you put a steel mesh around the candle, with holes of just the right size, the metal would carry away the heat, and the candle's flame wouldn't ignite the surrounding air.

Davy built an oil lamp using carefully sized tubes to let air flow safely through. He also wrapped the flame with mesh, which let light through but absorbed heat. Even though there was a flame burning inside the lamp, it could be taken into an area full of explosive gas without setting off an explosion—and even better, any methane passing into the lamp would be gradually and safely burned off.

The lamp did more than prevent explosions—it also gave useful warnings. When the air contained methane, the lamp's flame jumped higher and turned blue at the top. If the amount of oxygen in the air started to drop, the flame would flicker and go out—that was a cue for miners to leave the area before they, too, were "snuffed out."

Davy presented his invention to an audience of scientists at the Royal Society, along with a scientific paper on the subject. The gentlemen-scientists thought his work was brilliant and gave him applause, praise, and a medal.

Davy was pleased at this recognition of his scientific and inventing genius. He was not so pleased a few weeks later when George Stephenson, a famous engineer whose steam engines were used in many mines, produced his own new safety lamp, one that looked suspiciously similar to Davy's. It was obvious to Davy what had happened: Stephenson, the engineer businessman, had copied Davy's invention and was trying to pass it off as his own.

Accusations flew back and forth. Eventually, a team of experts investigated the claims. Surprisingly, it turned out that Stephenson had also carried out practical experiments and really had invented his own lamp completely independently of Davy. The two inventors had looked at the same problem and come up with the same solution at the same time. The idea was, literally, "in the air."

The two men never patched up their differences over the lamps. Davy, who got most of the credit for the lamp, didn't believe the experts. He spent the rest of his life convinced that Stephenson was a thief.

For many decades, the "Davy lamp" was celebrated in polite society as a groundbreaking invention and a breakthrough in workplace safety. But the reality under the ground was a little different.

The miners liked the fact that the colour of the lamp's flame could warn them of poisonous or explosive gas. With that added security, the miners (or, more likely, their bosses) were willing to risk working in parts of the mines that had previously been written off as too dangerous to enter.

Worse still, the miners found the brightness of the safety lamps was inadequate—sure, the steel mesh around the flame was good at preventing explosions, but it also blocked much of the light. So, they worked around the problem, using the lamp for safety warnings and carrying a candle as a light to work by.

It was an early example of what is now called "risk compensation"—when changes are made that protect us from dangers, we tend to adjust our behaviours, taking more chances. It's why ABS braking didn't lead to a reduction in collisions.

The introduction of the safety lamp led to a surge of mine explosions, and more miners died than ever. The safety lamp was a good idea, but sometimes safety can be dangerous.

A RAFT OF LIFE PRESERVERS

The nineteenth century was a golden age for patents, and the inventions reflect the obsessions and fears of the time. People often travelled by boat or ship, but many had never learned to swim, so drowning was a special fear. Many patents from that period are for discreet flotation aids.

In 1874, August Roos of New York patented a "life-preserving trunk." Although made of steel, its lid sealed air inside and made it buoyant. It could be used as a regular travel accessory, but if the ship was sinking, escaping passengers could strap themselves to the trunk then jump into the water, where the floating box would serve as a swimming aid.

This assumes, of course, that the swimmer didn't bang their head against the sides of the metal trunk as they jumped with it into the water. It also assumes the trunk had been packed with lightweight materials. If it contained heavier objects, it would certainly sink to the bottom, dragging the swimmer down with it.

Of course, when a ship is sinking, you might not have time to find your trunk and drag it to the deck. In 1842, William Shecut of New York proposed a better alternative—a life-preserving stool. It looked like the simple, four-legged seats often found on the deck of a ship, but two hollow, metal floats were hidden underneath. Pull a strap and springs would flip the floats to each side of the stool like the arms of a chair. You could then sit comfortably in the water and wait to be rescued.

Although it would probably have worked as a flotation device, there seems no need to attach the floats to the stool. Why not simply use the space under the seat to store a couple of life-preserver vests? Whatever floats your boat . . . when someone sinks your boat.

Another American inventor, Herbert E. Crosswell of Rhode Island, came up with another odd, life-saving stool. His version was a regular wooden stool with some modifications—it had a small air compartment under the seat, and the wood in the legs was hollow. It seems a lot of effort to solve the problem of turning an object that already floats into an object that floats slightly better. On the plus side, if a person fell overboard, the passengers could have fun throwing wooden stools at them.

LONG-TERM LIFE JACKET

Crossing the ocean was a risky business in the 1800s. The cork life jackets of that era might have prevented drowning, but wearers would be cold and wet. If they didn't die of exposure, they still risked death from starvation or thirst while they waited to be rescued.

In 1876, an inventor named Traugott Beck, from Newark, New Jersey, turned his mind to this problem. His solution was a new type of life jacket, made from waterproofed canvas, reinforced with telescoping metal hoops. Once the occupant had climbed inside it, they'd resemble a cross between the Michelin man and a large ice-cream cone—the suit completely enclosed the castaway as they bobbed in the ocean waves.

The cone floated low in the water, with only the occupant's head showing at the top. If the waves were too choppy, a domed canopy could be closed to keep the interior dry. A little snorkel tube would allow breathing even in choppy seas.

The suit had a window so the wearer could watch the sea for approaching ships. Unfortunately, there seems to be no way to wave or signal, and the low profile of the floating cone would make it difficult to spot on the waves.

A bizarre artist's impression of the suit from the May 1877 issue of *Scientific American* shows a bearded man using the device, his head bobbing above the waves as he walks along the ocean floor. How did his ship manage to sink in such a shallow sea?

The inventor claimed that the space inside the suit would contain enough fresh water and dehydrated food to last a person for a month. He doesn't mention how occupants are supposed to relieve themselves—it would be the first creative challenge for the user. In any case, a person would have a few weeks to enjoy the open ocean before starving to death.

Inventors are often loners, and like many inventors in the water safety category, Beck went overboard trying to create a comfortable, self-sufficient life jacket for a solitary survivor. Of course, a much simpler solution is to have a little trust in other people and equip the ship with life rafts. They are easier to use, easier for rescuers to find, and more comfortable. They can also rescue more than one person at a time.

YOUR OWN ATOMIC SUBMARINE

Inventions for saving shipwreck victims and preventing drowning include many odd ideas, but one of the wildest came from a French woman named Paule Christel in 1966. Her "universal swimmer" claims to offer everyone "the joys of the sea"—especially those who can't swim.

The patent is actually for a bizarre combination raft and submarine. It has a small cabin where a single occupant lies flat on their stomach, just like a real swimmer, and operates car-style controls. A set of four long paddles on the sides of the machine allow it to swim underwater like a giant, robotic turtle. Don't feel too bad if this doesn't make sense to you: the entire patent is as incoherent as it is enthusiastic.

The machine's power comes from a "silent atomic propulsion engine," housed in a small compartment below the operator. No details of this extraordinary motor are shown or described; however, the patent does explain that, in the event of the engine failing, backup is available from a second, identical atomic propulsion engine, adjacent to the first.

When the device is floating on the surface of the ocean, the lower section of the swimmer can act as a raft. Bizarrely, a set of long struts unfolds between the base of the raft and the cabin, lifting the operator above the ocean's surface, so they can come out in the open and won't be splashed by any annoying waves—although positioned at the top of a small tower, floating in a choppy sea, it's likely they would feel the effects of seasickness.

The inventor made bold claims for the device: it would end congestion at sea and reduce deaths by drowning. In a shipwreck, a single swimming device (per passenger, that is) would do a far better job than clumsy lifeboats. Maritime tragedies would become a thing of the past.

Although Christel was more interested in saving life than destroying it, she was also pragmatic and suggested that her swimming machine could also be a powerful tool in wartime. Armies of atomic-powered swimmers might attack an enemy by sea, invading an enemy's shores. The mechanized swimmers could also drill holes in enemy ships, perhaps creating a few of those maritime tragedies she was previously trying to eliminate.

In peacetime, the universal swimmer could also be used for fun. She foresaw a sporting competition where people would compete to cross the English Channel in her nuclear-powered vessels.

Accompanying pictures, poorly drawn, show what appears to be a huge device, operated by a pilot who looks like a hybrid of middle-aged human and seal. A French flag flutters behind the pilot, and a collection of tiny oxygen balloons are arranged in front. Although the technical details of the patent are confused and hard to interpret, these atomic submarines would certainly be challenging devices to build, and probably far more expensive than the ships that were supposed to carry them as lifeboats. Christel's requirements are fastidious: "The device must be made of stainless materials and, above all, be fireproof and waterproof. It must be light, but stable to allow it to be easily thrown through the access door of a ship in distress."

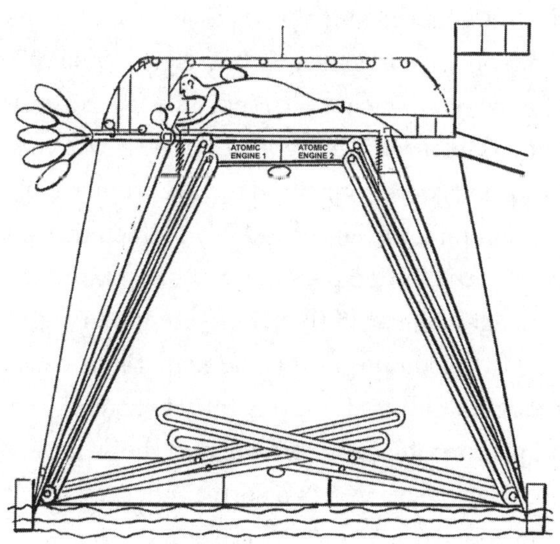

Ever hopeful, she ends by proposing laws that will drum up sales: "This device would also be mandatory for sailors sailing or working on the seas."

Sadly, no laws were passed to support her vision, and the joys of the sea, which the universal swimmer might have brought, remained a far-off dream.

SAFETY COFFIN

In the 1700s and 1800s, many people had a deep-seated fear of being buried alive. The fear even has a name—taphephobia. Fictional stories described the horror of the person who suffered a fever, lost consciousness, then awoke to find themselves in a coffin, buried under the ground.

Did this happen often? No. But that didn't stop a veritable army of inventors from coming up with systems to save the non-deceased.

A British minister named Robert Robinson designed a coffin for himself, and he was placed in it when he died in 1791. The coffin had a piece of glass on the front, so relatives could go into the mausoleum at regular intervals and check that the minister was still dead. He was.

A more common type of "safety coffin" used a signal, allowing the person in the coffin to communicate with those six feet over. One German inventor, Dr. Johann Gottfried Taberger, designed a system where the dead person's arms and legs were connected to ropes running up to a bell on the surface. If a person underground suddenly started thrashing about, the bell would ring and the buried person (or the vampire they had become) could be dug up again. It's said that the system was quite sensitive, and changes caused by decomposition could sometimes cause false alarms. The bells would certainly be a tempting target for morbid practical jokes.

A German priest named Pessler worried that a small bell might not be easily heard. His invention was a cord connected to the main church bells. The waking dead would wake the town.

In 1868, Franz Vester, an inventor from New Jersey, came up with a hybrid design; his safety coffin used both a bell and a window. A rope in the coffin sounded a bell at the top, sending a signal to anyone walking in the graveyard, so they could quickly rescue the revived burial victim. Vester's system also had a shaft running from the ground down into the coffin, which allowed passersby to peer down and ensure that the occupant was still dead. And if you were buried alive but awoke feeling spry and fit, the shaft had an interior ladder so you could climb to the surface yourself.

Vester's invention wasn't intended as a permanent fixture. After a few weeks, when there was no doubt that the person was quite definitely dead, the shaft would be removed and the hole filled in. The shaft could then be used for the burial of another safety-conscious customer.

It might be difficult to visually examine a person's condition by looking through a long, dark tube. A German priest named Beck offered a different idea. Each coffin would have a tube attached. The local priest would sniff it every day, and if it didn't smell like a dead person, the body would be dug up again.

In 1882, John Krichbaum of Youngstown, Ohio, devised a complex system where the mistakenly buried person could alert the living by turning a handle. The handle rotated a circular dial on the surface, changing the numbers it displayed. The occupant could also push upwards on the handle to open an air tube to the surface. It must have taken a calm mind to wait in a coffin until some mathematically inclined person noticed the change on a numeric display. You might get more attention by screaming for help up the air tube.

A German aristocrat, Duke Ferdinand of Brunswick, didn't trust others with his post-burial safety. His coffin was equipped with a lock that could be opened from the inside. He was buried with a key to the coffin and a key to the tomb that enclosed it.

Many of these safety systems provided an emergency air supply to the non-deceased. Some went further. Inventor Adolf Gutsmuth designed a safety coffin that also included a food supply. He demonstrated it on several occasions, having himself buried underground, while eating a full German dinner through a feeding tube.

This kind of demonstration could be risky, though. A Russian count named Karnice-Karnicki showed off his safety coffin by burying one of his assistants alive. The assistant was supposed to signal when they wanted to be brought up, but the signalling system didn't function, and the terrified assistant remained buried for much longer than planned. Fortunately, the emergency air supply worked.

Although safety coffins were mostly a phenomenon of past centuries, inventors still come up with new ones, using intercoms and electronic sensors.

In all the time these complex devices have existed, the number of people saved from live burial has been precisely zero.

CHAPTER 4

ALL THE COMBINATIONS

MANY inventions take two previously existing ideas and combine them in a new way. It's called a combination invention, and some are very creative. For example, in the 1980s, a British inventor named Trevor Baylis wanted to make a radio that could be used by people living in remote areas of Africa where there was no electricity or batteries. He combined a radio with a clockwork mechanism to generate electricity. The mechanism wasn't new and neither was the radio, but the combination was novel and useful.

But other combination inventions are downright strange.

WATTS'S SUITCASE BATH

In 1876, Ethelbert Watts thought of combining a bathtub with a large suitcase to make a portable bathtub "to afford travellers in places where such conveniences are wanting the luxury or comfort of bodily ablution."

Once you'd checked into your hotel room and unpacked your clothes from the case, there was no need to wait your turn to use the shared bath at the end of the hall. Instead, you could unfold the case into a large, waterproof box. You could then fill it with water and enjoy a relaxing suitcase bath.

But since the room didn't already have a bath or plumbing, where did the water come from to fill this suitcase bathtub? And once you had finished your bath, where did you put all that water? We assume the answer to both questions was the shared bath at the end of the hall.

GRATER GARBAGE MOUSETRAP

An inventor named Robert Gardiner, from Hamilton, Ontario, believed in giving consumers value for money. His 1897 invention looked like a cylindrical metal trash can, with a removable lid. You could use it to store your groceries.

There were holes and slits in the side. You could use these to grate or shred food.

Stand it on end, and it was a wonderful flytrap.

Put it on its side and insert a small flap in the lid, and it became a great mousetrap.

It would probably be a good idea to put a label on the container, so people could know whether it was currently holding trapped mice, dead flies, groceries, or freshly shredded cheese.

We're sorry we've come back to mousetraps again, but it does seem to be an obsession for the amateur inventor.

MULTI-PURPOSE MATCHBOX

You might have encountered the creativity training game where you're given an everyday object and asked to imagine possible new ways it might be used.

We're guessing that a similar process was responsible for an 1890 invention by Henry Brandt. The invention is a small, rectangular box.

The top of the box was padded and could be used as a pincushion.

The box lid opened, and the box could be used to store matches. There's a striker on the side.

Finally, you could also open the bottom of the box, prop up one end, and place a few crumbs in the back corner. If a mouse went under the box, the box would fall on top of it, sealing it in. The user can then drop the box into water to drown the mouse. The inventor points out that the pincushion is removable, so drowning the mouse won't soak the fabric, if that's your first concern.

Again, a mousetrap. And again, we're sorry.

BEAT YOUR PLOWSHARES INTO GUNS

In 1862, inventors C. M. French and W. H. Fancher were concerned about the safety of America's farmers who might be attacked by the Indigenous population or Confederate soldiers while plowing their fields.

They invented a combination plow and cannon. Anyone foolish enough to try attacking a lone farmer would face a blast of shock and awe from the cannon.

The cannon's muzzle faced forward on the plow, so it seems like the design carries a fair risk of the farmer shooting his own oxen or horses at close range from the rear, which is probably more encouragement than even the most stubborn animal needs.

HAMMER HAND SLING

It's hard to strike a hammer upwards, but miners might need to do just that. That's why, in 1899, Henry Schepers of British Columbia invented a new hammer—a combination of a hammer and a leather sling.

The tool consisted of a leather wrist strap for a hammer, but instead of being attached to the end of the handle, it was fastened to the head of the hammer. Schepers wanted to ensure that the working miner could avoid the "laborious effort of continuously grasping the handle of the hammer." With the strap around his wrist, he could place a chisel on the underside of a rock, twirl his hammer-on-a-sling, and bring it flying upwards onto the head of the chisel.

It sounds like one of those inventions that resulted from Schepers own experiments in the workplace. "That's an unusual technique you've got there, Henry. You should patent it." "By jingo, I think I'll do just that!"

It also sounds like a workplace accident waiting to happen. It would certainly require considerable skill to aim a heavy weight on a leather strap without smashing the hammer onto your hand or arm, or letting it fly back into your face.

It also introduces the possibility of the whirling sling-hammer coming loose, flying through the air, and striking another miner's skull.

But actual mining safety didn't seem to be an overriding concern in those days. A picture accompanying the patent shows a middle-aged miner, wearing overalls and a cravat, preparing to take a swing at a rock overhang. Nothing protects his eyes from rock chips, he has street shoes on his feet, and the only thing between his head and falling rocks is a fedora with a burning candle tucked into the hat band.

TOBACCO TABLET

One of the more revolting combinations was invented by Francis Schwartz, a British man living in Brooklyn in 1921. It seems he had adopted the (mostly) American habit of chewing tobacco. But he was dissatisfied with the way the product was sold: the tobacco was fresh when he opened a packet, but by the time he'd worked his way to the bottom, most of what was left was dust, unfit for chewing, and the rest had also deteriorated from contact with the air.

Schwartz then came up with a combination invention that proposed a bold rethink of the way chewing tobacco was sold. Instead of tobacco being sold loose, in a paper packet, he thought it should be put into a more modern and efficient form—compressed into a rectangular tablet, so not a particle of tobacco dust could escape. And instead of wrapping the entire pack in paper, each tablet of precious tobacco would be sealed inside a coating that would protect it from air and moisture. Chewing gum would be ideal.

His invention was a product shaped like a chocolate bar, made of mint chewing gum, with each square filled with chewing tobacco.

When a young tobacco fiend wanted to chew on some fresh tobacco, they could just break off a piece, pop it in their mouth, bite into the mint-flavoured gum, and enjoy the tobacco-leaf filling.

We don't know if this idea ever took off, but for those who don't find tobacco bad enough in its existing forms, this tobacco-and-chewing-gum-flavoured taste sensation raises the disgust factor to new levels. And if you thought getting someone's gum on your clothes was bad, try gum infused with a hairy layer of chewed tobacco.

A TWO-FOOT SPITTOON

George Babcock Clarke had a practical solution to cold feet in 1865. It was a foot warmer that sat on the floor. It was a cylinder with space inside for a burner. It ran on various fuels—the owner could either place hot coals in one set of chambers, or light two oil lamps in a different set. The heat warmed the round plate on top of the device. Put your frozen feet on the plate, and they would soon feel warm and toasty.

You know how some people have a good idea but don't know when to stop? Clarke was one of those. Having designed a sensible foot warmer, he clearly felt it needed something extra, so he added a spittoon in the middle. When your feet are warm and toasty, you don't want to walk away to spit.

The combination seems to diminish the relaxing possibilities of the foot warmer. Now, instead of one or two people sitting back and warming their feet, they would need to remain constantly alert for passing strangers with a mouth full of chewing tobacco who needed to spit surplus brown saliva into the spittoon.

The spittoon's hole is unusually small, so any spitter would need a very precise aim to get his tobacco-laden drool into the hole rather than onto the hot plate. It seems likely that it would end up all over the feet of other frozen travellers.

Even if the brown expectoration happened to land on target, it would be in the middle of what is essentially a small oven, inevitably sending the smell of tobacco-flavoured saliva through the room. But perhaps people were used to that in 1865.

WALKING ON WATER

Not many inventions allow the user to perform miracles, but that was the promise of a 1914 patent from Marout Yegwartian of Manchester in England. His invention was an "apparatus for walking on water."

Disappointingly, his apparatus appears to be nothing more than two small kayaks, one on each foot. Progressing on the water involved sliding one foot forward, then the other, like cross-country skiing. A series of pivoted paddles keep the boats moving in one direction, but it sounds like it might not have been an easy contraption to use, because he adds: "For balancing and propelling purposes the user is provided with a pole or paddle."

So now "walking on water" means standing in two boats and paddling.

The wording of the patent hints that the inventor might have encountered other problems in his tests. He suggests tying the boats together with elastic to prevent "undue separation of the feet."

We don't know if Yetwartian's invention was ever sold to the public, but it seems likely that any customers would soon return to the seller, soaking wet and complaining. Then again, that might allow the inventor to place his device under a different miraculous category—turning water into whine.

CHAPTER 5

YOU STOLE MY IDEA!

INVENTORS are a notoriously secretive lot, fiercely protective of their creations, and often paranoid that others are out to steal their latest idea.

In most cases, the inventor's fear is unjustified—frankly, most ideas aren't worth stealing, and some you couldn't give away.

But every now and then an idea comes along that is simple, elegant, and very valuable. When an invention is that desirable, others may indeed try to steal it, and the fights that follow can be brutal.

ERASING A PATENT

In 1922, Roger Watkins ran the Strathmore Printing Company in Aurora, Illinois.

One fall day, a man named Frank Rawlings came to see him.

"I'm an inventor," said Rawlings, "and I've come up with an amazing new idea for a writing tablet."

It was not much to look at—a cardboard frame covered with a layer of black wax. Above it was a sheet of white celluloid. Rawlings used a wooden stylus to write on the top layer. The celluloid stuck to the black wax where the stylus had pressed, making his writing visible. Rawlings then lifted a section of the frame, separating the white sheet on top from the black wax below. Instantly, the writing vanished without trace.

Watkins was impressed by this magic trick and could immediately see its value. This simple tablet could be used as an erasable notepad. It was lighter and cheaper than a chalkboard. It would save money on single-use paper pads. If the top layer was printed with a permanent grid, it could be used for common business forms, like timesheets. Yes, there was potential for profit here.

"If you're interested," said Rawlings, "I can sell you the rights."

Watkins certainly was interested, but he felt like he was being rushed, and he didn't like that. Something about this deal made Watkins uneasy. He asked Rawlings for a little time to think it over. Rawlings agreed.

Watkins didn't get to think about it for long, though. That night, he was awakened by a phone call. Frank Rawlings was on the line. He told Watkins that he had been thrown in jail. He had no money, and now he was desperate—he offered to sign over all rights to his invention to Roger Watkins, if Watkins would pay his bail.

What was a businessman to do? The inventor's misfortune was a great opportunity for Watkins, and this turn of events seemed too good an opportunity to pass up. Roger Watkins paid the bail and took ownership of the patent on Rawlings's invention.

Rawlings left town and Roger Watkins quickly started producing the device he called the "Magic Slate" for businesses, turning out millions of the useful tablets.

If Watkins's treatment of the penniless inventor seems heartless, you'll be pleased to learn it wasn't such a good deal after all.

Rawlings was a fraud. The only things he invented were lies.

The real inventor of the erasable writing pad was a German named Hermann Deutsch. He had patented the device in Berlin a couple of years earlier, around 1920. It was manufactured there as the "Printator." A cardboard frame surrounded the writing area like a window frame. When you wanted to erase the sheet, you pulled the writing area up toward the top of the frame, where a strip placed between the layers wiped all the markings clean. The area behind the sliding panel could be printed with an advertising message that would be revealed each time the tablet was erased. It was perfect as a promotional item, and people were fascinated by it. Sigmund Freud wrote about it, considering it a metaphor for the way the mind could create and erase information. The Printator company showed its product all over the world, seeking distributors.

Frank Rawlings had seen advertisements for the Printator and signed up as a salesman. He then turned his job into a con, finding Americans who had never heard of this European gadget, then fraudulently claiming he owned the patent to the product and wanted to sell it. Roger Watkins was just one of his victims.

However, Rawlings's story about needing bail money was perfectly true. While he had been scamming another victim in Kansas, he had started a relationship with a sixteen-year-old girl. Although Rawlings was already married, he ran off with her to Illinois. The authorities caught up with him shortly after he'd met Watkins.

Roger Watkins wasn't aware that he had bought a bogus patent when he started producing and selling his Magic Slates. It wasn't long before he received a letter from the German company's American lawyers.

Suddenly, Watkins faced a huge problem. He couldn't keep producing the Magic Slates without legal trouble, but after investing heavily in the new product, he didn't want to just walk away either.

He decided to sidestep the problem. He made deals with other inventors who had come up with similar types of devices, buying those rights to justify his claim that his product didn't violate the Printator patent.

He didn't want to compete head-to-head with Printator, but perhaps there were other uses for the Magic Slate. He had noticed that his kids liked playing with the device, so instead of targeting businesses, he sold the product as an educational toy for children.

As it turned out, Watkins's scramble to avoid a financial disaster turned out to be an excellent business strategy. The child-oriented Magic Slate was a major success, and the toy remained popular for more than seventy years.

Strangely, although the Magic Slate was reimagined as a toy, it ended up having more serious uses. During the Cold War, staff at the US Embassy in Moscow used Magic Slates to communicate information that might otherwise be intercepted by microphone bugs.

The original German product also underwent a transformation. With the rise of the Nazis during the 1930s, its use changed from business to military. The erasable pads became standard equipment for the Luftwaffe. Bomber pilots and navigators used them to make their calculations.

It was all very bad news for the inventor of the Printator device, Hermann Deutsch. That's because he and the business's other owners were Jews. The Nazis took the company from them and gave ownership to one of its non-Jewish employees.

It's not known what happened to the inventor after that, but the lack of records is itself ominous. Like the writing on his remarkable product, his history, and his very existence, may simply have been "erased."

INVENTING AN INVENTOR

In the 1700s, everyone was excited about telescopes. If you could afford one of these high-tech gadgets, you could use it to gaze at stars or examine distant ships far out at sea.

But despite the price, the images these telescopes produced were poor—objects often had a rainbow-coloured halo. This sort of distortion is called chromatic aberration, and nobody could figure out how to get rid of it. Most people figured it was just one of the limitations of using a lens. That's what Isaac Newton had said, and he (literally) wrote the book on optics.

John Dollond was a silk weaver with a lively mind. He thought maybe there was a way to solve the problem. After all, the human eye uses a lens, but it doesn't produce a coloured distortion. Dollond realized that the different transparent substances in the eye were bending light in different and complementary ways—one layer spread the colours in one direction, creating a spectral blur, and another bent it back again, fixing the problem. Perhaps it was possible to do the same thing with glass.

He played around with different types of glass. One type was called flint glass, because it was once made from melted flint. It was expensive, and it bent light quite strongly. Used on decorative glassware, it gave a glittering, jewel-like effect.

Another, cheaper type was called crown glass, because it was usually blown into a bubble or "crown," then spun outwards to make a round sheet. Lenses made with this glass didn't bend light as much as the flint glass.

Dollond tried combining flint glass and crown glass lenses in different ways. Finally, he found a combination that cancelled out the chromatic aberration and produced a very clear image. He published his results in 1758. The scientific community was highly impressed by his work and gave him a medal for the achievement, then invited him to join the ranks of the Royal Society.

Dollond patented his invention and started selling top-quality telescopes. They were used for everyday viewing by sailors, military officers, and landowners. He ran an optical company with his eldest son, Peter, and he was soon making so much from his telescopes that he could afford to give up the silk-making business. Dollond became the most famous lens-maker in London and was given the role of official optician to King George III.

So, a happy ending? Not entirely.

Dollond's telescopes were so much better than those made by his competitors that the other opticians had a hard time staying in business. So, they ignored his patent and copied his telescope design.

John Dollond's son Peter could see what was going on, and he was outraged at this flagrant theft. This needed to be stopped! But his father disagreed. He didn't want the fuss of a court case. He had a good life, and he wanted to enjoy it.

Sadly, his good life wasn't a long life. While studying the stars one night, John Dollond died suddenly from a stroke. Now it was just Peter Dollond running the business on his own, and he was determined to protect his father's intellectual property, even if his father hadn't. He launched a series of court cases against other telescope makers.

The legal battle went well at first. Dollond won several cases, and his competitors started getting nervous. It was just a matter of time before they were all shut down. They didn't have a legal case, so they formed a conspiracy. Sure, the facts were stacked against them, but that was no problem: they just needed to change the facts.

When Peter Dollond took the other opticians to court, challenging them with infringing his father's patent, they offered a surprising defence.

"Your father's patent isn't valid," they said. "He didn't invent the special lenses in this telescope. It's an old idea. It's been around for years."

The opticians spun a good tale. They claimed that a man named Chester Moore Hall had designed an identical lens to Dollond's, but thirty years earlier. Hall had been a wealthy lawyer and landowner with an interest in science. They all claimed to remember dear old Chester Moore Hall, the well-known reclusive telescope inventor.

Their claim didn't sound very likely. Finding a solution to this lens problem had been a major scientific challenge for John Dollond. If this man Hall had made such an achievement all those decades ago, why didn't he tell anyone?

The opticians had an answer for that. By their account, Chester Moore Hall had been shy and secretive by nature. He had told almost nobody about his groundbreaking idea. He was not commercially minded. He had built two telescopes but had given them away as private gifts to friends. He had only wanted to solve a difficult scientific problem, not make money.

But this explanation raised other problems. If Hall kept so quiet about his invention, how did the opticians know about his work?

They had an answer for that, too. They claimed that Mr. Hall had tried to keep his secrets by using the services of two different opticians to grind lenses. Each was given the job of grinding a single lens, but they weren't told it would be joined to a matching lens being crafted at a different shop. However, by a strange and remarkable coincidence, the two opticians had subcontracted the work to the same craftsman, a man named Bass, who saw both lenses at once. When Bass realized how the lenses fitted together, he knew at once what Hall had achieved, and the work of this discreet and modest gentleman became well known within the optician community.

The opticians claimed that John Dollond was one of a few people who knew about Hall's work and used Hall's methods, but that Dollond then had the audacity to patent it as his own invention and get rich off another man's scientific brilliance.

Peter Dollond was furious as he heard these lies. It was obvious to him that the opticians were trying to undermine Dollond's patent so they could stay in this profitable business themselves. They planned to steal the Dollond invention by inventing a different inventor.

Of course, the conspirators had very little evidence to back up their story. They were unable to show any telescopes that had been made before Dollond started selling his. But they brought in a series of witnesses who swore that, yes, that was just the way it happened. Unsurprisingly, most of the witnesses were people who would benefit financially if Dollond lost his case.

The opticians probably thought they had a slam-dunk case and an appealing story—surely the upper-class judges would prefer an aristocratic and modest gentleman-scientist to a mercenary silk maker who built telescopes for profit.

But they were in for a disappointment. In a ruling that is still studied by lawyers today, the judge said that even if the aristocratic Chester Moore Hall had made this important breakthrough, as the opticians had claimed, it didn't matter, because Hall hadn't done anything with it. The way the opticians told it, Hall had kept the details of a valuable invention locked up in his writing desk. If that was true, then as far as the rest of the world was concerned, he might as well not have invented anything at all.

The judge then turned his focus on Dollond, and the image was crystal clear. The inventor had obviously done a great deal of original work on lenses, he had published information about his findings, and he had produced and sold functional telescopes. Whatever the truth about Hall, Dollond had the first legitimate claim, and his patent remained valid.

Peter Dollond kept the patent and was able to stop the other opticians from stealing his father's ideas. His optical company became the optical firm of Dollond and Aitchison, which stayed in business from 1750 to 2015, when it was finally absorbed into Boots Opticians.

A RAZOR BLADE FIGHT

If you've ever looked at a "classic" rectangular razor blade, you've probably noticed the strange space down its centre, a gap that looks like an ornate candlestick. It makes sense that a blade might need a hole to lock it in place, but why did it need to be such a complex pattern?

In fact, the complex shape isn't necessary for any technical reason. It's all that remains of a ruthless patent battle between manufacturers in the 1920s.

It started in 1904, when an inventor named King Camp Gillette invented a new style of safety razor. It had two unusual features—the blade was sharp on both sides, so customers got twice as much blade for their money, and it was also disposable, so you never needed to sharpen it.

Gillette's business got a big boost when military regulations required all soldiers to have their own shaving kit. The company provided a neat kit packed in a military-style tin and sold more than a million of them. A generation of young men had been introduced to this new style of shaving, and after that, Gillette became the market leader in razors.

One thing quickly became clear from Gillette's business—making money from razors wasn't about selling the razors but about selling replacement blades. When Gillette's original patent ran out in 1921, other companies rushed to grab a share of the disposable razor-blade business.

One inventor, Henry J. Gaisman, sold his own Gillette-style razors, under the brand name Probak. Gaisman also made his own improved blades, more flexible than Gillette's and cheaper to make.

The original Gillette blade had three round holes to keep the blade in place. It was a simple design and worked perfectly well inside a razor, but with the patent running out, Gillette needed to make sure that customers would keep buying Gillette blades for their Gillette razors, so they made plans to give their new generation of blades a new custom shape. In addition to the three holes on the original blade, they added a horizontal slot, which fitted over a matching bar on their new razors. The new style Gillette blade would fit the old Gillette razors as well as the newer ones, but if you tried to put a Probak blade into a new Gillette razor, the bar would get in the way.

Somehow, Gaisman got wind of Gillette's plan and made a plan of his own—a devious one. He went back to one of his old blade patents and made some unusual "amendments" to it. Inventors are permitted to change an old patent in order to make minor corrections and clarifications, but Gaisman managed to slip in a several significant changes to the wording, claiming that his patent covered any extra non-circular hole or slot in the blade. That included a long horizontal slot—like the one Gillette was working on.

Just as Gillette was about to unveil the exciting new blade that was supposed to protect the company's business, Probak beat them to the punch, announcing the new Probak blade. It looked basically the same as the one Gillette was about to announce, and the new Probak blades were a perfect fit for all Gillette's new razors—although Probak had also added some extra slots and posts, preventing Gillette blades from being used in a Probak razor.

Gillette wanted to accuse Probak of stealing its ideas, but Probak's boss Gaisman claimed it was the other way around—Gillette had stolen Gaisman's ideas, and thanks to his carefully altered patents, he had the documents to prove it.

The battle between the companies went on for a while, but finally, the pressure on Gillette proved too much. Gillette was forced into a merger with Gaisman's Probak company. Although Gillette kept its name, the deal was expensive for the company and very profitable for Gaisman. When King Camp Gillette died in 1932, Gaisman replaced him as the head of the company.

The distinctive shape of a razor blade was fixed after that. The pattern of holes remained, like the fossilized memory of that battle between two inventors and their companies. Their products may have been safety razors, but the competition between them was cutthroat.

NEEDLE AND THREAT

Marie Louise Killick was a British audio engineer in the 1940s, and she wasn't impressed by the sound quality that came out of record players.

The big problem was the record player's stylus or "needle." Record players used a steel needle, and they didn't last long. Needles were sold by the tin, with a couple of hundred needles in each container. Some guidelines said you were supposed to change the needle every time you changed the record. All needles damaged the record a little, but a worn needle would sink further into the grooves of the record, doing permanent damage. Some companies offered heavy-duty needles, tipped with ultra-hard metals—those were good for twenty-five records, but were also quicker to damage the record. Some consumers bought soft needles, made of wood or cactus spines—they did less damage to the record but didn't sound good.

Killick realized that the record-playing world wanted a needle that was both durable and gentle on the records. In 1945, she invented one and patented it. The tip was made of sapphire—a material that is nearly as hard as diamond, but much cheaper. It would last far longer than any steel needle. But the clever part of her new needle was its shape: instead of being formed into a hard, record-gouging spike, it was carefully shaped with tapered sides and a blunt tip. When Killick's sapphire needle was used on a record, the needle made contact only with the edges of the record groove, not the bottom, where dirt tended to accumulate. Spreading out the area of contact meant minimal damage to the record. It also reduced "surface noise" caused by dirt in the groove.

Her invention wasn't just theoretical—Killick had a solid engineering background, and she knew exactly how to make these needles in bulk. She filed a patent for a machine that could carve the new stylus from a tiny piece of ultra-hard sapphire.

The sound quality from the new stylus was fantastic—better than anything that had gone before. It was a breakthrough in recording technology, and everyone could see it. Her "Sapphox" brand needles sold fast.

The record company Decca tested the new stylus and was blown away. They said her sapphire needles were the best on the market. They offered her the enormous sum of £750,000 if she would sell them the rights—to put this sum in perspective, at that time in England, a large, detached suburban house in a good neighbourhood might have cost about £4,000.

A woman inventor had spotted a product that sorely needed to be improved and found a way to do it. It was the right product at the right time, and it seemed like the inventor would reap the rewards of her ingenuity. What could go wrong?

As you can probably guess, the answer is "plenty."

Killick said no to Decca's offer. She figured she would be better off making the new needles herself.

With the benefit of hindsight, Killick would have been better off taking the money, but at the time, she had good reasons for declining the payout. If she controlled manufacturing, she would own a business that could provide a steady income for years. On the other hand, if she sold the rights for a huge flat fee, she would have to deal with the UK's heavy taxation at the time: ninety-five percent of the money she received would immediately vanish in tax.

So Killick forged on, building her sapphire needle business. She was confident in the quality of her product and trusted that her rights would be protected by her patent and the legal system that stood behind it.

But she had a product that big companies were desperate to get. Some of them played dirty. A giant electronic equipment company, Pye Ltd., ignored her patent and started producing a needle just like hers. Her sales started declining.

Killick fought to protect her patent in court. Pye fought back hard and made sure it was a slow, expensive legal process for Killick, at a time when her business was suffering because of Pye's pirated products.

She also ran into a different, unexpected problem—she had hired a professional patent agent to take care of registering and renewing her patent. Incredibly, he had forgotten to renew it. It should have been possible to correct the mistake, but the patent controller was strangely reluctant to do it. Finally, after long delays, he admitted that Killick's patent agent was an old friend of his. If he reinstated the patent, the agent's mistake would be a matter of public record, and his friend could be sued. The poor fellow would be ruined. It didn't seem to worry the patent controller that the incompetence of the patent agent and the delays in fixing it were costing Marie Killick a fortune in lost sales.

Killick kept pushing and got the patent error sorted out. She also refused to back down on the court case against Pye. She eventually won the case and was awarded five million pounds. It was a big victory—at least on paper. Pye found ways to delay making any payments. While she waited, Killick found herself struggling financially.

The battle was not only legal and financial. According to those around Killick, the company also used intimidation to beat her down. She received threatening phone calls. Intruders tried to break into her home.

At one point, she was even grabbed on the street and forcibly taken to a mental institution. It turned out that her abductors were police officers—someone somewhere had faked a doctor's medical report and had her committed.

The dirty tactics worked. Killick finally declared bankruptcy and disappeared from public view. She didn't see any of the money she had been awarded by the courts. Given more time she might eventually have triumphed, but the stress of the battle had taken a heavy toll on her mental and physical health, and she died in 1964.

Like the sapphire needle she invented, she had kept playing for a long time but was eventually worn down.

LET ME BAG THAT FOR YOU

Many women inventors have seen their ideas stolen by men, but some have managed to fight back successfully. An early example was an inventor named Margaret Knight.

In 1867, Knight got a job at the Columbia Paper Bag Company. She was not impressed by the machine-made bags the company produced. They were folded like giant envelopes, and if you put something large inside them, they split apart. If you stood them on end, they fell over.

In those days, you could buy a flat-bottomed paper bag. It was big and roomy, much better for holding groceries, and it stood on its base. The problem was that the only way to make such a bag was to glue it together by hand, so it was expensive.

Although Knight hadn't had much formal education, she had the brain of an inventor, and she had already come up with creative improvements to existing machines, including mechanical looms and machines that handled paper. She was sure she could design a machine that would take a piece of paper and fold it into a flat-bottomed bag.

She worked at the idea and built a prototype from wood. It worked exactly as she had intended. This invention was worth patenting. Unfortunately, the American patent rules at that time required her to submit a working iron model to apply for the patent. She had to go to a machinist's shop to have this version made.

While the device was being crafted, a machinist named Charles Annan visited the shop. He was impressed by her invention and could see the potential in it.

Finally, the iron prototype was ready, and Knight took it to the patent office to apply for the patent. She was shocked to discover that a patent had already been filed—under the name Charles Annan. He'd liked her idea so much that he'd stolen it.

Knight took Annan to court. She had a mass of technical drawings showing how she had developed the design. She also brought in witnesses who supported her version of events. She won her case and got her patent. She sold her design to a paper bag company and lived off the income. Her invention earned her an award from Queen Victoria.

Knight kept applying her skills and came up with dozens of other inventions. She was always more interested in the creative work than the business side of things. When she had invented something new, she would sell the patent and live off the royalties. She was still working on new inventions into her old age.

For women inventors, in particular, it sometimes feels like "no good idea goes unpunished." But Knight managed to buck that trend. While Knight never became super rich from her work, she lived her life doing things she loved.

IT TURNED OUT ALL WHITE

Another woman inventor who beat the system was Bette Nesmith. She has two claims to fame: she was the inventor of Liquid Paper, and she was the mother of musician Mike Nesmith, from the pop group the Monkees.

Bette Nesmith divorced her husband when her little monkee was four years old, and she supported herself and her son by working as a secretary at a Texas bank. At some point in the 1950s, the bank invested in new IBM electric typewriters. Typing was not Nesmith's strongest skill, and the new typewriters made things worse—not only did the sensitive keys make it easier to produce typos, but the carbon-film ribbon was difficult to erase. A small mistake could mean having to retype an entire letter.

Nesmith was an amateur artist. When artists make a mistake, they don't have to erase it—it's often possible to paint over it and keep going. Perhaps what worked in art would also work in typing. She mixed up a batch of white paint in her kitchen blender, poured it into nail varnish bottles, and took some to work. The next time she made a mistake while typing, she just painted over the error with a fine brush. When the paint was dry, she could type a correction on top of the paint. In most cases, the correction looked perfect.

Other secretaries noticed her technique. They were all facing the same issues trying to correct errors on the new typewriters, and they asked Nesmith if she could mix some of her correcting paint for them. She was happy to do it, and she realized that this technique might be a product she could take to market.

Nesmith had been using white poster paint for her corrections, but she knew it would be better if the paint was more durable and dried faster. She asked for help from her son's high school chemistry teacher (Walter White-out?). Soon she had a formula that worked even better for typists.

At first, selling the bottles of correction fluid was only a minor side-hustle for Nesmith. The extra money helped, but she still had a hard time supporting herself and her son. Things turned suddenly worse in the late fifties, when she mistakenly typed the wrong ending on a letter for work. Instead of signing off with her boss's name, she typed the name of her home business, "The Mistake Out Company." Her boss was not amused and fired her.

Rather than seeking another office job, Nesmith threw all her energies into her correction fluid product. She pitched big manufacturers like

IBM, offering to sell them the idea. Surprisingly, they showed no interest—they were more interested in selling typewriters to businesses than they were in solving the problems faced by the typists.

So, Nesmith sold the correction fluid herself. She changed the name of her product from Mistake Out to Liquid Paper. It took a few years for the company to make any profits, but by the mid-sixties, things were starting to look good, and the same companies who had turned down the rights to Liquid Paper were now paying to buy thousands of bottles. She and her son both made it big around the same time. While the Liquid Paper Corporation was fast growing into a multi-million-dollar business, with factories in Toronto and Brussels, Mike Nesmith was making a good living as a member of the Monkees.

Dozens of other manufacturers copied her idea and produced similar products. But in the face of this stiff competition, Nesmith remained scrupulous and careful. She didn't focus on selling more, but on finding ways to make her deceptively simple product a better experience for the people who used it in their workday. Its formula was a closely guarded secret, but it worked well. Even when the market became flooded with imitation products, her version remained the market leader.

Other problems were more personal. Nesmith remarried, and her second husband helped her expand the company, but when their marriage fell apart, he tried to push her out of the company and take control of it himself. She held her ground and fought back, managing to hold her stake in the company.

She eventually sold her share to the Gillette company for nearly $50 million. When she died in 1980, half her fortune went to her musician son, and the other half went to charities supporting single women in business and the arts.

CHAPTER 6

OFFICE SUPPLIES

BETTE Nesmith's success with Liquid Paper shows that an inventor can get rich selling the right office products.

Businesses are about making a profit. They're willing to spend money on a newly invented product if it saves money. Office sales drove the success of ballpoint pens, dry erase markers, and calculators, as well as software like spreadsheets and word processors.

But coming up with that profitable office invention isn't always easy.

THE QUIRKS OF QWERTY

Although most people don't use actual typewriters anymore, the skill of typing is still going strong. All over the world, people tap away on computer keyboards that are a close copy of the old mechanical keyboards on a typewriter, their electronic keys spaced, sized, and arranged in pretty much the same way as typewriters from 150 years ago.

The most common English-language keyboard has the QWERTY layout—named for the first six keys in the top row of letters. It's an odd arrangement, and you might have heard the explanation that the grouping was chosen to slow down typists. Early typewriters were unreliable and prone to jamming if two keys were pressed in quick succession. It's claimed that the inventors chose a more confusing layout of letters to slow down speedy users. Today, more than a century and a half later, billions of keyboards are still stuck with a keyboard layout that was first chosen for its inefficiency.

It's a great story about the human resistance to change, but sadly it doesn't seem to be true. In fact, the jumbled letters were an inventor's genuine attempt to make his keyboard work faster.

The inventor was named Christopher Latham Sholes, and he designed the first really successful typewriter. His earliest models were nothing like any typewriter we'd recognize today. He built his machines with a piano-style keyboard, including black and white keys with a weighted mechanism. The letters were laid out in alphabetical order.

The keyboards were slow to use, so he looked for ways to make them faster, more efficient, and more "scientific"—this was the trendy word for inventors in the 1800s, especially among those looking at new forms of communication, like shorthand systems and telegraph codes.

Sholes worked to optimize his keyboard. To minimize finger movement, he reduced the size of the keys and arranged them on three rows instead of two. The weighted keys were slow to rebound, so he replaced them with keys on metal springs that instantly snapped back into place after each keypress.

He also thought about how changes to keyboard layout might make typing faster and less tiring. He reasoned that the effort should be spread between all the fingers of each hand, so each finger did its fair share of the work. He started moving keys around to spread the load.

He also tried to avoid having long stretches where only one hand was active. Common pairs of letters—TH, SH, NG—were placed on opposite sides of the keyboard, so the typist used first one hand, then the other.

He also worked hard to avoid having a single finger type too many different keys in a row. There are some exceptions—words like "yummy" (typed with the first finger of the right hand) or "ceded" (the middle finger of the left hand)—but not many.

Far from being inefficient, the QWERTY keyboard does a decent job of distributing activity between both hands. Most letters ended up being scattered from the original alphabetical layout, but a few remained together—if you look at the middle row of a modern keyboard, you can still see remnants of the order Sholes started with—the FGH and JKL.

QWERTY was the layout Sholes had settled on when he finally sold his typewriter design to Remington, a firearms manufacturer trying to make a move into less dangerous machines. The typewriter was well received, and people quickly got used to the new keyboard and developed methods to "touch-type" with it.

Despite his success, Sholes was a restless character and never seemed quite happy with his own keyboard design. Even after the typewriter had taken off, with its now-familiar layout, Sholes was fiddling with newer, better layouts. In 1896, he filed a patent for one he considered better. It had all the vowels on the middle row, under the right hand, and the most common consonants right above them.

Whether Sholes's newest keyboard was better or not, it didn't matter. Nobody was interested. By then, most typewriter manufacturers had adopted the QWERTY layout, and the layout was supported by a growing industry of typing schools. People weren't interested in starting over with a new system.

Since the first QWERTY typewriter appeared, many inventors have spent time designing newer and more "scientific" keyboards. But when alternative systems have been carefully compared or tested in typing speed competitions, the speed differences between QWERTY and alternate layouts have proved surprisingly small. If it ain't broke . . .

BETTER, FASTER, CHEAPER

There's a saying among engineers—"Better, faster, cheaper—pick two." But what if you could invent a product that checked all three categories? Then you'd have a big success on your hands, right?

The answer is, "It depends."

Remington manufactured the first successful typewriters, but their early models had a few issues. The company had started as a gun manufacturer, then moved to sewing machines. They built their typewriter the same way they built their sewing machine—big and heavy. The typewriter was designed to sit on a desk and stay there. Their early machines had other issues, too—the typebars jammed if you pressed two keys at once, and the route the paper took through the machine meant you couldn't see what you'd typed until you were finished. Remington had a few competitors, but their typewriters were just as huge and cumbersome.

George Blickensderfer ran a delivery business. He spent hours travelling back and forth on trains and found himself wishing he had a typewriter small enough to take with him, so he could work on the train. He started thinking about ways to reduce the weight of an office typewriter.

Blickensderfer was a natural inventor. Even as a boy, he tried to build a flying machine. It didn't work—or, if it did, he kept very quiet about it. He also had an inventing role model in the family: his uncle, a man named John C. Zachos, had invented a successful courtroom stenotype machine. Blickensderfer wasn't intimidated by the idea of designing a new typewriter.

He demonstrated his machine in 1893, at an exhibition in Chicago. The hall had many typewriter manufacturers showing their heavy machines. In the middle of it all, a young woman typed away at sixty words a minute on a revolutionary new typewriter. Gradually, visitors left the other exhibitors and gathered around the Blickensderfer booth. People couldn't believe what they were seeing. Blickensderfer hadn't just reduced the weight of the typewriter—he'd redesigned it from the ground up. The result was an engineering masterpiece.

The new machine was so stripped-down that it looked like the skeleton of a typewriter that somebody had forgotten to put back together. Three rows of keys fed under a slim paper roller. The main mechanism was all contained in a tiny compartment in the centre. It was tiny and toy-like, and yet it was clearly rugged—it typed well and fast. And unlike the other machines in the hall, you could actually see the words you were typing as you typed them.

The mechanism was deceptively simple, using only 250 parts—a tenth the number found in most typewriters. That meant there was less to go wrong. It also meant the typewriter could be built more cheaply than its competitors—the Blickensderfer was being offered at less than half the price of a Remington. It was unbelievably light. It could be easily lifted with one hand or carried in a case.

But the most amazing feature was that his machine had no typebars to get jammed. Instead, the letter forms were attached to a little cylinder. Press a key and the cylinder turned to the correct letter, rolled against an ink pad, then placed a perfectly formed character on the paper. It all happened at lightning speed.

Putting the type on a cylinder meant you could easily change the machine's font. Replace one cylinder with another and you could type italics or script-style letters. Want to type German or Greek characters? Just swap out the cylinder with one for your chosen language. This was a killer feature in 1893, decades ahead of its time—in fact, it wouldn't be seen again until the 1960s, when IBM introduced its "revolutionary" Selectric typewriter, with its "golf ball" printing element.

The new typewriter was a sensation, and people immediately started placing orders. Europeans loved it—especially the ability to use multinational characters. It did well in North America, too, as a portable machine for reporters.

Blickensderfer had a manufacturing hit on his hands. He sold thousands of the new machines. For a time, it was the best-selling typewriter in the world.

Better, faster, *and* cheaper!

This is probably the point where Blickensderfer should have started hiring others—engineers and marketing experts—to help him run his growing business and expand it in a strategic way. He was way ahead of the competition, and perhaps he could have stayed ahead. Instead, he stayed a hands-on boss, bursting with creative ideas.

Around 1900, electricity was an exciting new technology. Blickensderfer figured it was the key to a new generation of typewriters. A typist could work faster and with less effort if a motor assisted in moving the mechanism. He started work on the new idea, and in 1901 he showed his electric typewriter to the world. Again, it was an astonishing breakthrough. The machine was a desk model. It was slim and elegant. A typist could touch the keys lightly and produce perfect letters.

The electric model included another revolutionary innovation. On earlier typewriters, when the typist reached the end of a line, they had to pull on a lever to move the carriage back to the left and advance the paper. But on Blickensderfer's electric model, you could just press a special key with your right pinkie and the carriage would reset on its own, like magic. It was the first "return key" and that key (and its placement on the right) is still with us today.

The new typewriter should have been a hit, but George Blickensderfer had failed to consider a couple of problems with this machine. The first was the price—at $125, it cost twenty-five percent more than a manual machine, and that was too expensive for most customers. (There goes cheaper!) A bigger problem was that electricity was still a new thing—most people didn't have it yet, and the electrical systems weren't standardized from region to region. Even in areas that were wired up, the electricity wasn't available twenty-four hours a day: some power companies assumed electricity would be used for lights and only turned it on at night.

Then there was the keyboard. Blickensderfer had been unimpressed by the QWERTY keyboard used on Remington typewriters. Instead, he created his own improved "scientific" keyboard, putting all the common keys on the bottom row (spelling out DHIATENSOR). Less common letters were placed above them along the middle row, and uncommon letters were at the top. A shift key and a second modifier key meant that each of the regular keys could produce not just two but three characters—lowercase, uppercase, and a number or punctuation mark.

His keyboard was certainly clever and made efficient use of keyboard space. Perhaps it was also faster, as he claimed. But ignoring QWERTY was a bad move. The QWERTY layout had become universal on typewriters, and when typists were used to it, they found it hard to switch to Blickensderfer's layout. Blickensderfer was confident that his keyboard was better. It seemed irrational that anyone should want a worse one. He pushed the "scientific" keyboard on his electric models, too, hoping it would catch on. He grudgingly offered a QWERTY option—although not a recommended one.

So, the new machine had several strikes against it. Nobody had been clamouring for an electric typewriter, and although it worked well, very few people bought it. (If you have one today, it's valuable.)

The electric typewriter was far ahead of its time, an engineering marvel, and a spectacular failure in the marketplace.

Disappointed, Blickensderfer went back to making improvements to his manual typewriters. He made new improvements. He used an aluminum frame on his portable machines. The model was advertised as the "five-pound secretary" and remains one of the lightest portable typewriters ever made. But as the years went by, his business gradually became less profitable.

The inventor was pouring all his creativity into his product line, but other manufacturers had entire teams of engineers to work on theirs. The typewriters from companies like Remington and Underwood steadily improved, becoming cheaper, lighter, and faster. Several companies produced portable typewriters. They weren't quite as portable as the "Blick," but they were sturdy and worked well—and they had keyboard layouts people were used to. As the years went by, Blickensderfer's machines no longer seemed the trailblazers they had been when they first came out.

But things really turned bad for Blickensderfer when World War I began. He had depended on European sales, but now Europe was at war, and nobody was buying. He only stayed in business by making machine gun parts for European customers. (Ironically, in shifting from typewriters to guns, he had gone in the opposite direction from Remington, a company that had moved from guns to typewriters.)

George Blickensderfer died before the war was over. He had been a truly brilliant engineer and the company's guiding force, but although his company had factories and salespeople, it was too much of a one-man show at the top level. When George was gone, nobody could fill his shoes. His heirs didn't know how to run a typewriter business, so they sold the company's designs to other manufacturers. Eventually, the Blickensderfer typewriter was purchased by its old rival Remington. In the 1920s, Remington tried marketing the "Rem-Blick" as a low-cost typewriter for students. But by then, the design was starting to look very old-fashioned. The last models were sold in the late 1920s as children's typewriters.

You can have better, faster, and cheaper—but not for long.

CHEAPER, SLOWER, WORSER

Just because someone has invented a typewriter, that doesn't mean you can't invent a worse one. A few years after the first keyboard typewriter was invented, manufacturers introduced the index typewriter. This was a machine where you used a mechanism to select the next letter, then pressed a lever to punch that letter onto the paper. If it sounds slow, that's because it was.

The cheapest and simplest models (known appropriately as simplex typewriters) involved turning a dial to select each letter. This was a cumbersome operation, better suited to making short labels than typing correspondence.

A more sophisticated, and somewhat faster, version used a stylus, which could move left and right, up and down over a printed grid of letters. As you moved the stylus, a mechanism would line up the correct letter. Tapping a bar would stamp the letter onto the page.

Although these stylus-based typewriters were much slower than a machine with a keyboard, they were also much cheaper, and they did work. If an office didn't write too many letters, and workers had time to burn, it could serve as a cheap alternative to a keyboard typewriter. The devices were sold as a low-budget option for years, until finally the cost of typewriters with keyboards became low enough that there was no need for this kind of compromise.

Even then, index typewriters didn't entirely disappear but continued on as children's toys. The simplex models were often offered in newspaper ads as a "real typewriter" (sporting a non-functional printed keyboard) at a suspiciously low price.

A similar mechanism was used more successfully in handheld machines for making embossed plastic labels, most famously by the Dymo brand.

THE TYPING GLOVE

Some inventions are so obviously bad that it's hard to know if the inventor was serious or just trying to con the public.

The first modern typewriter, the Remington No. 1, hit the market in 1874, and within a couple of decades, the machines had appeared in every office. But they remained expensive and complex machines. The cost of buying just one typewriter could pay the salary of the typist for half a year.

In 1891, one inventor, a Mr. Carey, decided it had all gone too far. The high price of the typewriter was outrageous and, as he saw it, its complexity was unnecessary.

Get rid of all the complicated machinery, said Carey. He replaced the typewriter roller and case with a simple wooden box holding a stack of paper. In place of the typewriter's confusing mass of machinery, he offered a simple leather glove, with rubber letters glued to the fingers, and an ink pad on the palm.

Carey arranged the letters of his typing glove using a weird version of the QWERTY layout. The letters ran in the opposite direction to the typewriter keyboard—the little finger contained the letters QWERT running from fingertip to palm. The ring finger continued the top typewriter row, YUIOP, and started on the second row with AS. The middle finger held the remaining letters of the middle typewriter row, DFGHJKL, and the index finger contained the typewriter's bottom row, ZXCVBNM.

The rubber letters stuck to the thumb were a series of useful shortcuts—AND, THE, OF, ING, TION. For only $2.50, the buyer received one typing glove, and a small wooden box to hold the paper.

The invention was steadily improved. A later version of the "Glovegraph" gave the typist a glove on each hand—one for lowercase, one for capitals.

Typing with the glove was a process of elegantly tapping the page with different parts of the finger, occasionally clenching a fist to put ink back on the little rubber letters.

In this way, said Carey, "the human hand takes the place of all the complicated mechanism of key machines, and guided by the eye, the letters can be placed accurately in position with even greater rapidity than is attained in the most improved key machines." It was touch-typing without the need for clumsy keys.

It's reported that he demonstrated his Glovegraph to the members of the London Shorthand Writers' Association. Sadly, there are no reports of anyone buying the device.

In reality, it would be almost impossible for a glove-typist to write rapidly or neatly. There was nothing to keep the letters in a straight line, and each cautious press of an inky rubber letter against the page carried a high risk of including the edges of its upper and lower neighbours. Worse, brushing the palm against the page would leave large black marks from the ink pad.

Carey's claims were ridiculous, and so was his promotional material: an accompanying photograph shows a sad-faced woman wearing a pair of industrial leather gloves that seem to be sized for a silverback gorilla.

The Glovegraph was cruelly—but fairly—mocked by magazines of the day.

MONEY SOMETIMES GROWS ON TREES

The public will often recognize the value of a clever gadget like a typewriter, but when a creative mind comes up with a simpler, bigger, and much more profitable idea, people often don't recognize it at all.

Canadian inventor Charles Fenerty came up with one of those big ideas. In the 1830s, he was a teenager living near Sackville, Nova Scotia. The Fenerty family was in the lumber business. Charles Fenerty had different dreams: he was a poet, as well as a part-time inventor.

He loved walking and thinking, and one of the problems Fenerty thought about was paper. In the 1830s, it was expensive. Paper was made from the soft fibres like cotton or hemp. The cotton usually came from old rags, but the world was seeing an explosion of office work and publishing, and there weren't enough rags to meet the demand for paper.

During his walks through the woods, Fenerty examined wasp nests. Wasps made their paper nests by chewing wood into a soft pulp, and he wondered if humans could do the same thing mechanically.

He spent a few years working on the problem. He found a way to make the wood fibre soft, scraping it off the surface of the wood in fluffy clumps, then turning it into pulp. After many experiments, he figured out how to use these wood fibres to produce high-quality sheets of paper.

In 1844, Fenerty wrote a letter to Nova Scotia's leading newspaper, the *Acadian Reporter*. He included a sample of the paper he had made, which was just as good as anything made from cotton fibre. He suggested that the most common trees in Canada's vast forests—trees like fir and spruce and poplar—were a particularly good source of wood fibre. He encouraged others to try the process and see what they could do with it.

That's as far as Fenerty took it. Having solved the problem, he returned to his poetry and other work.

The letter was published. It didn't generate much interest. Nobody contacted him. Nobody investigated the method he had described.

The following year, a German inventor named Friedrich Keller patented a similar process. Keller got recognition for his discovery, and within a few decades, making paper from wood pulp had become a massive industry around the world.

As for Fenerty, he later spent a few years in Australia, trying to get rich in the Australian gold rush. Perhaps he might have found a bigger bonanza if he had patented his breakthrough wood pulp invention.

But Fenerty wasn't really the one who missed out. After all, he knew what he was doing when he published his newspaper letter. He made a choice to give his valuable invention to the world for free. In modern terms it was a billion-dollar idea, and anyone who had read his piece in the newspaper could have run with it and made a fortune. But sometimes the world just isn't that bright. Even with a winning idea placed right in front of them, in a location filled with forests and lumber mills, there were no takers, and the opportunity slid by.

"Shoulda, coulda, wood-a . . ."

PULP FRICTION

If you want to make money from an idea, you need to believe in it, patent it, and push it out into the world. At least, many inventors believe that, but it doesn't necessarily do the trick.

While Charles Fenerty was trying to turn wood into paper in Nova Scotia, Friedrich Keller had been working on the same problem in Germany.

Keller was also a young man—just five years older than Fenerty—and, like Fenerty, he came from an ordinary working family—his father was a weaver. Keller wanted to be an inventor and engineer, and he jotted down notes whenever a new idea entered his busy brain. While in his twenties, he read a book about papermaking that discussed the theoretical possibility of making paper from wood. Keller was intrigued by the concept and set to work on a machine that could grind up wood to make suitable fibres.

By the time he was twenty-eight, he had cracked the problem: with the aid of his wood-grinding machine, he could make paper from wood. But he had gone as far as he could go—taking the technology further would require money. He excitedly wrote to the German government, explaining what he'd accomplished and requesting financial support for his important discovery. The response was disappointing. Nobody cared.

Both Friedrich Keller in Germany and Charles Fenerty in Canada had announced the same breakthrough discovery within a year of each other. The paper industry faced an overwhelming demand for paper at the time, and it should have been obvious that the new inventions would solve supply problems, but on both sides of the Atlantic, nobody seemed to appreciate what the inventor had done or the opportunities their inventions offered.

Fenerty's response was sanguine. He shrugged and moved on to other projects.

But Keller was much more committed to his invention than Fenerty and seems to have been more determined. He was certain he was onto something big, and when the government wouldn't put money into it, he looked for a paper manufacturer who would. He eventually connected with Heinrich Voelter, a man in the paper industry. Voelter loved Keller's ideas and paid him a small upfront fee for the rights. It was important to get the invention patented, and the two men did this together, under both their names.

Pretty soon, the new machine was being used to manufacture paper in Voelter's mills. Voelter didn't keep the technology to himself—he also sold the wood-grinding machines to other paper manufacturers across Germany and around the world. As the years went by, more and more publishers started using the inexpensive new paper, further increasing demand for the machines that made it.

Keller's belief in his invention and his determination to see it used had been the right decision, and it led to huge profits. Unfortunately, Keller didn't see any of them. That's because, although Keller was a co-owner of the patent, the deal he'd made with Voelter didn't give him any royalties—all he'd received was Voelter's upfront payment. Meanwhile, Voelter made big money from Keller's machine. He teamed up with other equipment manufacturers. He hired engineers to make further improvements to the technology. Voelter forgot all about Keller—he had no further need for Keller's involvement.

When it came time to renew the patent a few years later, Keller had so little money that he couldn't afford to pay for his share. Voelter renewed the patent on his own and became the sole owner of a patent, which was rapidly revolutionizing the world of publishing and really "made paper." As a man of wealth and influence, he even became a government minister.

It's a sad story, but Keller did eventually get some recognition.

Most people in the papermaking industry had assumed that the inventor of their key technology must have made a fortune from it. But when Keller was in his fifties, and living in poverty, the real story became widely known. People were appalled at how badly he had been treated. Other business owners organized a collection for Keller to show their gratitude. It didn't make Keller rich, but it was enough for him to buy a small house and support himself in his old age.

COPYING PRESS

In the 1700s, the Industrial Revolution was in full swing. Business was booming, and so was bureaucracy. People often wanted copies of documents—and that was a problem. There were no photocopiers in those days, so the only way to have a copy of a letter you'd sent was to write it twice.

James Watt was one of the big inventing names from that time. He had made a fortune manufacturing steam engines, but the business of making and selling the machines soon took over his life. He found himself overwhelmed with office work. He often wanted copies of letters and technical diagrams, and he was frustrated that he couldn't do it more easily.

In 1780, Watt came up with a way that any letter-writer could make a copy of their work.

The idea was simple. After you'd finished writing your letter in ink, you placed a piece of tracing paper (also called onionskin) on top of it. If you pressed both sheets together, quite hard, some of the ink from the letter would get transferred to the back of the tracing paper. Of course, the transferred text was reversed, but because the tracing paper was semi-transparent, you could read the words quite easily from the front side.

The invention needed a couple of refinements. The ink needed to stay wet long enough to be copied. Watt came up with his own special slow-drying ink formulas, adding ingredients like sugar to regular ink. It also helped if you made the tracing paper slightly damp before you pressed it down.

Pressing the sheets together evenly required a mechanical press. Watt was in his element with this problem. He loved machinery, and he came up with two solutions. With his first machine, you turned a hand crank and ran the papers through a pair of rollers. That worked well if you only needed to copy one letter at a time. A second machine, designed for heavy-duty office use, resembled the screw presses used in bookbinding. You could put in a whole stack of letters all at once, each with its own sheet of tracing paper, and the press would copy them all in one go.

The machine was a little messy to use. You could only use it with Watt's special ink, which stayed wet for a full day—so it was easy to smudge your neat handwriting. The ink for the copy all came from the original letter, so you could only make a single good copy. And moistening the tracing paper seems to have been a skill that took practice—if you didn't do it just right, parts of the copy would be faded or missing. Still, the copier was cheap to operate, and once you got the knack of it, it worked well—and there was no better way to copy letters.

Watt's company produced a portable version. It came in an elegant mahogany box that contained a pen, the special ink, a writing surface, a pad to wet the copy paper, and a roller mechanism to squeeze the papers together.

Benjamin Franklin loved the portable unit and bought one for George Washington, who used it for his presidential letters. The larger "copying presses" also caught on. The tracing paper was usually bound in a large "copying book" with numbered pages. Secretaries placed their letters between blank pages of the book, put it under the press, and turned a handle to squeeze it all together and copy the letters.

These devices were common items in offices for the next hundred years, until they were finally replaced by typewriters and carbon paper in the twentieth century.

Today, this method of making copies is almost totally forgotten, but people still remember who invented it. Ask anyone "Who invented the copying press?" and they will likely say "Watt?"

CARBON COPIES

Ralph Wedgwood came from a family of highly inventive potters. They had made a fortune finding ways to make affordable copies of expensive plates, vases, and teapots.

The Wedgwoods were friends with James Watt, who had recently produced the successful letter-copying system. Watt's copying system worked well but was quite expensive. So, following his family's tradition with pottery, Ralph Wedgwood did the same thing more affordably.

In 1806, Wedgwood revealed his invention: a special paper, coated with a mix of oil and black pigment. He called it "carbonated paper." Today we'd just call it carbon paper.

That seems like it might be an entire solution right there—people could stack two pieces of paper, with a sheet of carbon paper in the middle. If they wrote a handwritten letter on the top sheet of paper, they'd get a second copy below.

Unfortunately, it wasn't that simple. In the 1800s, most people were still writing with an old-fashioned ink pen. The nib might be a trimmed feather or perhaps a steel nib, but neither one makes enough of an impression to produce a carbon-paper copy.

However, Wedgwood had a way to deal with that problem. He'd noticed that James Watt's copying machine worked by printing a mirror-image copy onto the back of a piece of translucent paper that the writer could read from the front side. Wedgwood borrowed this idea. His copying system used a sandwich of sheets—on the top was a sheet of clear tracing paper. Below it was a special sheet of carbon paper, coated with black pigment on both sides. On the bottom layer was the regular paper that would be sent as a letter. As for the pen—you didn't use one. All the writing was done with a metal stylus.

When you wrote on the top layer, the carbon paper made marks on the back of the tracing paper, which you could easily read from the top side. The underside of the carbon paper also made clear black letters on the plain paper below. The system offered other advantages too—users didn't need to worry about spilling ink. The lack of ink also made it useful for visually impaired writers—with regular quill pens, they could never tell if the ink from the most recent dip had run out.

Wedgwood turned a good profit from his invention, but although it was well received at first, it didn't sell as well as he had hoped.

There were two problems. One of them was legal. Many business dealings required an original document, written in ink. With Wedgwood's system, there was no ink "original," only two copies made from Wedgwood's carbon paper.

The second problem had to do with the carbon paper he'd invented. It was made from pigment and oil. It printed well, but it smelled terrible. Thomas Jefferson tried Wedgwood's "manifold copier" and couldn't deal with the ungodly stench. He said: "The fetid smell of the copying paper would render a room pestiferous, if filled with presses of such papers."

Wedgwood's carbon paper would be put on hold for a few decades, until it was later carbon-copied and re-released for use in typewriters. By then, it had become single-sided and made with an ink that didn't stink up a room.

DUPLICATING PEN

Thomas Edison was one of the world's most famous inventors, with 1,093 patents to his name, but not all of Edison's ideas were successful, and many ideas he thought would be surefire hits turned out to be duds. Then again, some of his duds later turned into unexpected success stories.

Bureaucracy was exploding in Edison's day, and everyone needed paper copies. James Watt's copying press allowed people to make a single copy of a handwritten document, but what if you wanted dozens of copies, or hundreds? Today, this sort of duplication is easily done with a printer or photocopier, but in Edison's day, there was no way to do that—if you wanted multiple copies, you'd need the services of a professional printing company, and that was expensive.

Edison was sure there would be a demand for a short-run copying system, and in 1876, he thought he had found the answer: an electric pen. You wrote your letter on a piece of waxed paper. As you moved the pen, a thin needle inside it would move up and down, dozens of times per second, punching tiny holes in the paper.

When you were finished writing, you loaded your waxed paper into a holder, with a sheet of blank paper underneath. After adding a few coloured chemical solvents and squeezing them through the tiny holes with a roller, a copy of the original writing or drawing would appear on the blank paper. The results weren't great, but they were certainly readable, and one stencil could potentially produce a thousand copies, at the rate of four or five a minute.

There were a few issues, though. The pen's ergonomics were bad. Although the writing end was shaped like a pen, the upper half sported a large metal wheel, which spun at high speed, driving the needle rapidly up and down through the shaft of the pen. The pen was noisy, poorly balanced, and not comfortable to write with. The system posed other dangers, too—most offices didn't have electricity, so the electric pen's motor was powered by a large acid-filled battery sitting on the user's desk. Writing a document carried the risk of being stabbed with a fast-moving needle or splashed with battery acid.

Given the problems, maybe it shouldn't come as a surprise that the device wasn't the runaway success Edison had hoped.

He ended up licensing his copying system to another manufacturer. They dropped the idea of needle holes, but Edison's chemical copying system was adopted for the mimeograph copier, which was used for decades in offices and schools as a cheap alternative to photocopying.

As for the cumbersome needle pen, it morphed into the electric tattoo gun used by tattoo artists. Instead of punching tiny holes in paper, it punches holes in skin. We're guessing even Edison's fertile mind wouldn't have thought of that use.

CHAPTER 7

SOUNDS AND PICTURES

INVENTING a new recording medium can transform the world. Today, we can listen to music and watch films because of breakthroughs in recording sound and colour images.

But innovation doesn't always mean quick profits, and the technological advances that launched Hollywood and the music industry got off to a slow and very rocky start.

Sometimes, the creators didn't even know what to do with their new inventions.

EDISON'S TALKING DOLL

One of Edison's earliest and most famous inventions was the phonograph—the first record player. The original was a cylinder covered in metal foil. A steel needle rested on the metal, attached to a small horn. If you rotated the cylinder while shouting into the horn, the vibrations of your voice were recorded as waveforms in the foil. Reverse the process, so the grooves vibrated the needle, and the recorded sounds played through the horn. The first versions didn't require any electronics—it was a purely mechanical process.

In 1877, people were amazed by this technology, but they wondered what it was good for. Edison wondered the same thing. The phonograph seemed like a solution in search of a problem. The sound quality wasn't good enough for music. Edison thought it might have potential as a dictating machine, but nobody had been asking for one. Most people viewed the device as an ingenious gadget—a mere toy.

Then Edison had a brainwave. If the phonograph seemed like a toy, maybe it could become one—he'd use it to make a talking doll. He set to work creating a miniaturized, 18-centimetre version of his player, using a short cylinder that could hold a few seconds of speech.

One of the first things Edison had ever recorded on his phonograph had been his own voice reciting the nursery rhyme "Mary Had a Little Lamb." He now continued this pattern. Each doll would speak a nursery rhyme aloud.

It probably looked like a simple idea at first, but Edison ended up spending years trying to refine the details of the doll. The final version was a mix of materials. The body was made of tin and was just large enough to hold the player mechanism. The playback horn was hidden behind holes in the doll's chest. The arms and legs were jointed wood. The doll had a porcelain head, manufactured in Germany.

Each phonograph needed its own recorded cylinder. This was a bigger problem than it sounds, because Edison didn't have any way to duplicate cylinders: each doll's voice had to be individually recorded. A team of female workers spent their workdays in the factory, reciting nursery rhymes into a horn. They didn't know it, but they were the first recording artists in history.

In 1890, thirteen years after he had come up with the idea, the phonograph doll finally went on sale.

The press was excited about the new product and predicted great things from Edison's latest invention. The doll represented a new generation of technological toys. One reporter predicted: "Millions will be manufactured and toydom will be revolutionized."

In fact, Edison's doll was a disaster.

The first problem was the price. The toy was very expensive for the time. The cheapest doll was $10—the equivalent of about $300 today. A version with a more ornate dress cost more than twice as much. This was a doll that only the well-heeled could afford.

Then there was the mechanism, which wasn't designed with a child in mind. Making the doll talk involved turning a crank in the doll's body. It had to be wound at just the right speed, or the speech would sound too high or low in pitch. When the brief recording was finished, the child was expected to lift another lever to reset the needle.

The quality of the voices was another strike against the doll. No microphones were used to make the cylinders, so the overworked women recording the verses needed to talk very loudly, almost shouting, if their voices were to be captured. It didn't leave much room for nuance.

If the child turned the crank too fast, the shouting voice would become a piercing scream; if too slowly, it turned into a threatening growl. It sounded creepy.

The dolls came equipped with many different rhymes, including a morbid prayer popular in the nineteenth century, when child mortality rates were still high:

Now I lay me down to sleep,
I pray the Lord my soul to keep.
If I should die before I wake,
I pray the Lord my soul to take.

The prayer is supposed to be comforting, but when recited in a high-pitched scream, the doll's sinister voice made it seem as if it intended to accelerate the dying process.

Children and adults found the dolls' voices disturbing, and Edison himself described the sound as "exceedingly unpleasant."

The machinery was delicate, too. An accidental drop or a jolt could stop the player from working. Even if it kept operating, the wax cylinder soon became worn by the needle. The doll's voice then sounded both harsh and indistinct.

The doll quickly acquired a bad reputation. Not many people wanted to pay the high price, and the few who did were usually unhappy with the toy and returned it to the store.

After a few months, Edison realized his invention was a money-losing dud. He walked away from the project, and the remaining mechanisms were eventually destroyed.

Edison's doll was decades ahead of its time, but its failure likely had a chilling effect on other manufacturers who might have considered a talking doll. In fact, customers who wanted to buy one that actually worked would have to wait another seventy years, until 1960 when "Chatty Cathy," was released. Like Edison's doll, Chatty Cathy used a miniature record player, but with many improvements—it was durable, the string-powered mechanism played at a steady speed, the phrases were well chosen, and the voice sounded like a cute child, rather than a screaming demon from hell.

SWEET, SWEET MUSIC

Edison might have made his talking doll more appealing to kids if he had waited a few years and taken advantage of an invention from the Stollwerck company in Cologne, Germany. In 1903, they patented a new technique for recording records. Instead of recording on wax, they proposed using a material that the company happened to specialize in—chocolate.

The company promptly put the product on the market. Their record player was tiny—similar in size to the player inside Edison's doll. Its clockwork mechanism was made by Junghans, the famous German manufacturer of clocks and watches. The record was an 8-centimetre disk of chocolate, impressed with grooves that held the low-fi, high-calorie sounds.

The company experimented with other edible options. For a slightly more durable recording, the chocolate could be wrapped in a "groovy" metal foil for playing and unwrapped only when it was time to eat it.

Of course, the chocolate records wore down much faster than a record made of hard wax, but children didn't care. Any annoyance they felt about the rapid decline in audio quality vanished once the owner placed the recording to their lips and took a bite of the smooth, rich sounds. It meant that the device wasn't the disaster Edison's dolls had been. However, chocolate records weren't a huge success either, and Stollwerck eventually went back to their main chocolate business and abandoned their foray into "soundbites."

Very few of the chocolate records have survived, and the remaining players are now very rare collectors' items.

There have been other edible records. In fact, before the 1940s, most records were theoretically edible. Old records were usually made of shellac. The material comes from an insect, and the food-grade version is widely used in confectionery. When it's an ingredient, it's usually listed as "confectioner's glaze."

Edible records have occasionally been produced as novelty items. In 2018, Kellogg's collaborated with the Simon Cowell boy band PrettyMuch and came out with a chocolate Frosted Flakes record—a disk made from frozen Frosted Flakes that was covered with layers of milk and dark chocolate containing the recording. The record was just a one-off, but if the promotional videos are any indication, it played well, and it looked appetizing. It seems a great option for bands you don't like—instead of burning their records, you can crumble them up and enjoy them in a cereal bowl.

VICTORIAN HOLOGRAMS

When you mention photography from the 1800s, most people imagine grainy black-and-white images. But while photographers were snapping grey portraits of stiff Victorian families, an inventor came up with a way to create brilliant colour photographs. The idea amazed the world, but that didn't make it a success.

Gabriel Lippmann fit the stereotype for a science genius. As a kid growing up in Luxembourg, he'd been described as "inattentive," but he was great at math, and his skills took him a long way. By the 1870s, he was a professor of physics at France's most prestigious university, the Sorbonne.

One of the nice things about being a tenured professor is that you get paid for working on anything that interests you, and what really interested Lippman were the pretty colours on soap bubbles.

Liquid soap isn't colourful, so why do the bubbles produce those brilliant rainbow hues? The answer is that the surface of the soap bubble is very thin—a similar size to the wavelength of visible light. Viewed from the right angle, the light reflected from the back of this film interferes with the light reflected from the front, causing some colours to be muted, while others are boosted. The result is a spectacular range of hues. As liquid moves around the bubble, and some parts of the liquid film get thicker or thinner, the colours brighten or fade.

The colours produced this way don't have to be rainbow shades. The same effect can produce specific colours. The iridescent feathers of a blue jay or a peacock create colours in a similar way. The bright blues and greens of their plumage are produced by microscopic ridges on the feather, positioned so they suppress other colours and let only blue or green light shine out. You see the same effect on some flies and beetles.

Lippmann knew all this, and he was fascinated by it. He wondered if you could use this effect to capture colours in a photograph. Suppose, instead of a soap bubble, you used a very thin layer of liquid, containing the tiniest particles of light-sensitive chemicals—would it capture the colours? In the early 1890s, he tried it, putting a photographic emulsion on a thin sheet of glass, then resting it on a layer of liquid mercury, which would work as a mirror.

He took a few photographs of objects that would look good in colour—a bowl of oranges, a parrot, a stained-glass window. It worked amazingly well. When the film was developed, the colours were recorded in a single, thin layer of film, like a permanent soap bubble. When the glass plate was held up to the light, the colours blazed out. The film was equally sensitive to every colour, and the images were true to life and beautiful.

The method was wonderfully simple. The colours of an image were automatically captured in the same way they had entered the film—no colour adjustment was necessary. As long as the photograph had been taken and processed correctly, it "just worked." His photographs still exist, and the colours still shine.

The world was amazed by his discovery. Many photographers and scientists copied his methods, and Lippmann was awarded the Nobel Prize for Physics in 1908.

And yet, despite all the attention and acclaim, Lippmann's method didn't take over the world, and today even most professional photographers have never heard of him.

What went wrong?

Unfortunately, there were a few practical problems with Lippmann's method.

One big one was that the photographs needed an extremely high-resolution film, with a much finer grain than most photographs of the day. And exposing those tiny particles required much more light than most other photography—his first photos needed to be exposed for fifteen minutes. Later, he reduced that time to one minute, but that's still a long time to sit still and say "Chee-e-e-e-e-ese."

Then there was the problem of viewing the photographs. Forget leafing through a photo album—the only way you could see these photographs was to stand in a dark area, directly in front of the photograph, with a light directly behind it. It worked better if people looked at the photographs one at a time—if you were off to one side, the colours vanished and the image became a drab, grey shadow on a black-and-white film plate.

And perusing your snapshots also required the right atmosphere—literally. If there was too much humidity in the air, the gelatin in the film absorbed the moisture and became slightly thicker. The change was microscopic, but it was enough to alter the way the light waves interfered with each other. Changes in humidity could cause a temporary shift in the colours or even make them vanish completely.

The size of the photographs was also a problem. Lippmann's samples were about as big as a large postage stamp. Making them bigger meant exposing the image onto a larger glass plate—and that would require an even longer and slower exposure.

Finally, there was the problem of making copies. You couldn't. Because the colour information was stored at different levels of a thin layer of film and produced by internal reflection, the images were impossible to duplicate. You also couldn't project them onto a bigger screen or turn them into a paper photograph. It would have been like trying to project the colours of a soap bubble.

The upshot was that Lippman's award-winning breakthrough in photography was gradually abandoned and then forgotten.

His work wasn't entirely lost, though. Decades later, similar methods were used in a different way, not to store colours this time, but to produce 3-D holograms. And today, a few photographers still experiment with Lippmann's methods for colour photography, which work better using modern materials, although the pictures are still tiny and hard to view.

But his invention wasn't the breakthrough in colour photography that people were waiting for. In fact, producing good colour photographs would have to wait another forty years after Lippman's first photographs, ten years after his death. Making colour images that were easy to view would require a method of processing that was vastly more complicated.

NOT A BLACK-AND-WHITE PROBLEM

Rube Goldberg's cartoons were famous for showing complicated machines that performed a simple task in a very complex way. Some real-life inventions worked the same way. But sometimes a convoluted method is the only way to do things, and it can work pretty well.

Around 1915, two students met at the Riverdale School in the Bronx. Leopold Godowsky and Leopold Mannes shared a first name, and they soon discovered they also had a few other things in common. Both came from musical Jewish families (Godowsky later become brother-in-law to George Gershwin) and both had inherited a passion for music. They were also passionate about science. And they were also very interested in photography.

At the time, photographers around the world were still waiting and hoping for a way to take colour photographs in the same way they could make black-and-white ones. The best colour process available in those days used two colours, usually green and red. The results looked more like a tinted photograph than lifelike colour. Godowsky and Mannes went to see a new colour movie about the navy. The colours were terrible. They figured they could come up with something better.

The two Leopolds set to work on the problem, experimenting with different ways of filtering light and trying to come up with a practical, three-colour photographic process.

They came up with some good ideas, and their work was intriguing enough that Kodak hired them as full-time employees. The company had a big research laboratory where many top chemists were working on the same problem. The scientists had made useful discoveries involving dyes and bleaches—they were busily inventing methods that might someday be usable as part of a colour film process, but nobody had come close to producing a complete system that could be used commercially.

The other chemists were amused by Godowsky and Mannes. The pair lacked the deep scientific knowledge that some other employees had, and their colleagues referred to them as "those musicians," because as they did their chemical experiments, they would time the reactions by whistling pieces of classical music—they both knew how long various pieces of music ran, and this produced an accurate timing.

In 1930, the big-budget colour feature film *King of Jazz* hit the cinemas. It was a musical revue featuring the Paul Whiteman Orchestra. Since the orchestra's most famous performance was "Rhapsody in Blue," it was embarrassing that the film's two-colour technology made it impossible to reproduce the colour blue. Technicians did their best to give an impression of blue, but it was more of a rhapsody in grey.

Around the same time, Godowsky and Mannes were coming to the end of their contracts with Kodak, and they worried about their future. The Great Depression was getting worse, and the two oddball musicians hadn't made any major breakthroughs. But if they could somehow crack the problem of colour film, their future at the company would be assured.

Desperate to find a solution, they decided to try a shotgun approach. They said, "If we collect everything the people here know so far—all the steps people are sure of—can we put these methods together to make a process that works?" They tried it. They put together all the research they could find on the different dyes and processes that might produce a full-colour film, and they worked out a way that all these methods could be combined. At least, in theory.

But in practical terms, their method seemed completely ridiculous. The film would need to use many ultra-thin layers of material sandwiched together. Some layers in this microscopically thin sandwich would be sensitive to one of the primary colours, while others formed a thin barrier between layers. Manufacturing such a film would be a major challenge.

But making the film was easy compared to the nightmare of trying to develop it. The way the two Leopolds planned it, the film would need to be briefly exposed to chemicals so the liquid could seep partway into the film's emulsion coating, adding colour dye to each layer. They had to soak just enough to reach a certain layer of film. The emulsion on film is very thin, so for this method to work, the distance the liquid travelled had to be judged to within thousandths of a millimetre.

Imagine a sink full of water. Now imagine dipping the flat surface of a square kitchen towel into soapy water for just long enough that it makes the back of the towel wet but keeps the front side dry. Does that sound hard? Here's a second challenge. Wait until the towel is dry and dunk it again, in milk this time. This time you need to dip it for just long enough that there's milk on one side, the remainder of the soap in the middle, and the back of the towel is still dry.

That's the sort of solution "those musicians" had come up with—although much harder, because a kitchen towel is about a hundred times thicker than the emulsion on film.

What made their invention worse was that the developing process used around thirty different steps, and every one of them involved liquids penetrating to these precise depths, along with some difficult chemistry. Each step had to be carried out with exact timing and at precise temperatures—the slightest error would ruin the result. Even if everything worked, their method was painfully slow: it took several hours to develop each film, and there were several stages of washing and drying the film.

Their method was obviously completely crazy—the chemical equivalent of a Rube Goldberg machine. But it did have one thing going for it compared to the more elegant methods being researched at Kodak: this one worked. If you did it right, it could actually produce a beautiful colour photograph.

In the end, Kodak embraced it and went to market with this system. They built huge new plants that could carry out all the precise and complex chemical steps required on an industrial scale. They called the new film Kodachrome. Each box of film was sold with a mailer envelope, so it could be shipped to Kodak for the painstaking development process.

Kodachrome was a huge success, for both movie and still cameras. Within three years of its introduction, Kodak was selling more colour film than black and white.

Although the film was difficult to process, the colours were excellent. As it turned out, they were durable, too. Photographs and films taken on Kodachrome hold their colours and look good after many decades, whereas many later photo films (which were more chemically sophisticated and much easier to develop) tend to fade badly over time. Kodachrome had been introduced as a stopgap—a way to get colour photography out to the public while the company refined better methods—but incredibly, the film continued to be sold for nearly eighty years. It was not until 2009, well into the era of digital photography, that Kodak finally shut down the last of its Kodachrome developing plants. Developing Kodachrome film is now impossible.

CHAPTER 8

REINVENTING THE WHEEL

"DON'T reinvent the wheel," the old saying advises. But inventors can't resist.

Sometimes they were vindicated. Attempts to reinvent the wheel have produced some impressive achievements in cars, trucks, and high-speed trains.

They have also produced some impressive failures.

BOILING ROAD RAGE

Steam trains were a big deal when they were invented, and when gasoline-powered motor cars appeared on the scene, they were a success, too.

So why weren't there more steam-powered vehicles for the road? Aside from some oddball models like the Stanley Steamer, which used gasoline and kerosene to boil water, most steam vehicles ran on tracks.

In fact, steam-powered road carriages got off to a very promising start. They were a big improvement over everything that had gone before. The biggest pioneer of this technology made smart decisions and produced a good technology . . . and things still went horribly wrong.

The inventor's name was Goldsworthy Gurney (from the sound of it, he may have been part hobbit). He was a surgeon and gentleman-scientist, from Cornwall in England. That part of the country was home to several steam engine inventors, but those engines were mostly used in mines, and nobody had yet managed to produce a really good road vehicle. Gurney decided it was time to take steam engines out from under the ground and onto the highways.

Around 1825, he designed and patented an ambitious new steam carriage. It was a big vehicle—roughly the length of a modern full-sized pickup truck—and weighed about 2 tons. It was tall, too—the chassis and steam engine components ran along the underside, so everything else was raised up. Passengers needed ladders to climb aboard.

Seating was more like a bus than a car. Twelve passengers sat in the open air on four rows along the top of the vehicle. Six more could ride in covered comfort in a compartment in the middle. The carriage had six wheels—two small ones at the front for steering, a medium-sized pair in the middle, and two large drive wheels at the back.

The rear of the vehicle held a large boiler and steam engine. Gurney had made ingenious changes to earlier engines, making his engine smaller, more powerful, and less smoky.

Gurney wasn't sure if steam-powered wheels would give enough traction in every situation—for example, if a driver needed extra push getting up a hill—so he added a bizarre extra feature to his first version: a set of steam-powered legs!

It turned out that wheels provided all the hill-climbing power the vehicle needed, so the pusher legs were amputated from later models. Too bad. If things had gone another way, we might now be travelling around in Star Wars–style "walkers."

The driver didn't get a steering wheel. Instead, they controlled the direction of travel using a long rod or tiller which connected to the front wheels.

Gurney had created an impressive piece of engineering. His steam carriage could travel on regular roads at an average speed of 23 kilometres per hour—which may not sound much, but it was faster than any of the stagecoaches in the early 1800s—and that was just the relaxed cruising speed. Gurney had already tested the carriage at a top speed of

32 kilometres per hour and was playing with ways to get more power from the engine.

For a while, Gurney was optimistic: the steam carriage seemed to be catching on. Other engineers liked his ideas and built their own steam carriages, running other passenger services.

An engineer named Walter Hancock built a "steam omnibus"—a similar vehicle to Gurney's. Hancock named his bus the Enterprise, and it could boldly go where every man had been before, from one side of London to another. The driver of that vehicle used a steering wheel, like the ones in modern cars. Meanwhile, up in Scotland, another engineer named John Scott Russell began a regular steam-coach service between Glasgow and the town of Paisley, about 12 kilometres away.

Gurney was at the cutting edge of a revolution in transport. Thousands of passengers were being transported in the steam vehicles, and they enjoyed the speed and convenience of the new vehicles. The technology was steadily improving, producing more powerful engines. Roads were improving too. It looked as if he had hit the jackpot with his invention—prospects looked bright for fast steam carriages.

Unfortunately for Gurney and his fellow inventors, there were plenty of people who didn't like the new vehicles. A network of steam carriages posed an obvious threat to the existing network of horse-drawn stagecoaches. And it wasn't only the business owners who felt threatened—so were many workers, including stable hands, drivers, wheelwrights, blacksmiths, and innkeepers. On Gurney's maiden voyage, his steam carriage was attacked by an angry mob, who injured the driver—and burned their hands trying to attack the steam boiler.

Gurney also faced opposition from another group of engineers—the train barons who wanted to build railways. They were supported by the steel industry, who hoped to provide the rails. Although they were all steam engine enthusiasts like Gurney, they knew it would be hard to raise money to build a fast rail network in Britain if the country already had a "pretty fast" road network.

The opponents of steam carriages used every opportunity to undermine the technology. They put logs and rocks across roads; saboteurs damaged vehicles; and when minor accidents occurred, they were heavily publicized, making the public believe that the engines were likely to explode beneath them.

Gurney's enemies also used their powerful political connections against him—it led to legal changes intended to crush his business. One law placed a huge toll on steam-powered vehicles. Another law required that any motorized vehicle on British roads had to have a man walking ahead of it, carrying a red warning flag. The result was that steam coaches became prohibitively expensive to run, if "run" is the right word for a vehicle that can only travel at walking speed.

Gurney's ticket sales collapsed. Instead of making a modest profit, he found himself nearly £250,000 in debt and had to declare bankruptcy.

At heart, Gurney had always been more of a scientist and experimenter than a single-minded steam engine enthusiast. He knew he was beaten, so he gave up on his steam carriage and turned to different projects, where he had more success, including streetlights, firefighting equipment, and heating and ventilation systems for buildings.

But the result for travel technology was that a promising invention was quashed and suppressed. Steam-powered railways became profitable, while road travel remained stuck in the mud for decades.

BOILING OVER: STEAM CRASHES

As steam-powered road vehicles hissed to a halt in Britain, they had already done the same in France.

The very first "car" was a steam-powered vehicle from French inventor Nicolas-Joseph Cugnot in 1770. He didn't intend it for passengers, though. Cugnot was a military man, and he wanted to build a vehicle that could carry four people and pull a cannon. His work was sponsored by the French army.

It was not an elegant design. A slab of a chassis had two heavy ox-cart wheels at the back and a single wheel at the front, which did the steering. Instead of placing the steam engine at the back, which was how most later vehicles worked, Cugnot attached it to the front of his massive steam tricycle, so when the driver turned the steering wheel, he was also turning a huge metal boiler. This was not a stable arrangement, and the vehicle tended to tip over on uneven surfaces.

In 1771, Cugnot was driving a second version of his vehicle when he hit a wall, smashing right through it. The vehicle was so clumsy and awkward that the army finally gave up on steam-powered cars and went back to horses.

As for Cugnot, he goes down in history as the man who invented the world's first car and had the world's first car crash.

× × ×

Steam-powered road vehicles may have tanked in Europe, but inventors in North America continued experimenting with the technology. But when you were used to horse-drawn vehicles, the unexpected speed and power of steam could be dangerous.

An American inventor named Sylvester Roper built a steam engine and attached it to a typical American horse buggy–style frame, with a bench seat on top and four giant wheels. He would put on demonstrations of his new "steam buggy." According to posters, it was "pronounced by scientific men to be the most wonderful invention of modern times," and he claimed it could travel 240 kilometres a day.

Roper didn't limit himself to four-wheeled vehicles—he also constructed a steam-powered motorcycle. His "steam velocipede" could be "driven up any hill and will out speed any horse in the world."

In 1896, Roper demonstrated his latest steam velocipede in a race against top cyclists riding regular bicycles. The seventy-two-year-old Roper rode the steam bike and easily outran his fit young rivals, hitting speeds of 64 kilometres per hour. Then something went wrong—the vehicle became unstable, and Roper fell. He was dead by the time the others reached him.

Roper had built the world's first motorcycle and was the world's first person to die in a motorcycle crash.

In Stanstead, Quebec, a watchmaker named Henry Seth Taylor had seen a display of an American steam carriage, which was probably one of Roper's, at a travelling circus in 1864. Inspired by the demonstration, he started to build his own. By 1867, he had constructed a similar-looking vehicle—a buggy-style carriage powered by a two-cylinder, coal-fired steam engine. Taylor built the steam engine himself, using his metalwork skills. Like some earlier designs, it used a steering tiller rather than a steering wheel.

A crowd gathered to see the new buggy in action. Unfortunately, the vehicle quickly broke down, and the audience lost interest. Embarrassed, Taylor went home and spent a year refining the design. The following year, he rolled the improved version out at the Stanstead Agricultural Fair.

The vehicle ran smoothly for the second demonstration, so he took his buggy out onto a hilly road. It did a great job of climbing the hills.

Have you ever had the experience where you're not aware of a problem until it's staring you in the face? That's how it was for Taylor.

In designing his steam buggy, he'd been focused on making it go faster. But as he crested the hilltop, it suddenly occurred to Taylor that his buggy had a design flaw he'd overlooked—it had no brakes. He found himself going downhill, in an accelerating vehicle, with no way to stop.

Taylor jumped from the careening vehicle, and his buggy crashed into a creek. It was wrecked. He took the parts home, threw them in his barn in disgust, and turned his attention to steam yachts instead.

A century later, the remains of Taylor's historic car were still sitting in that Quebec barn. The wrecked vehicle was rediscovered and rebuilt.

As for Taylor himself, he goes down in history as the man who invented Canada's first car and—following the pattern of the other inventors here—had Canada's first car crash.

PUMP MY RIDE

The failure of steam carriages brought other technologies down with it.

A Scottish engineer named Robert Thomson was interested in both railways and steam carriages. A locomotive has hard steel wheels, which work well on smooth steel tracks. Steam carriages also had hard wheels; like the horse-drawn carriages they hoped to replace, they used a large, spoked wooden wheel with a steel rim. The wheels were built large, allowing them to roll over bumps and potholes, and the steel rims lasted a long time.

In 1846, Thompson wondered if putting an inflatable-rubber-type tire over the steam carriage wheel would work better. The experts said no—they claimed the rubber would create more friction than steel, making the carriage harder to move. But that was just a theory—nobody had bothered testing it.

Thomson was a skeptical and determined twenty-three-year-old, and he wanted to see the results for himself. He made some tires and, since he didn't have a steam carriage to experiment on, he fitted them to a regular horse-drawn carriage. It was a type known as a brougham—a four-wheeled carriage pulled by a single horse. He then took a second, identical brougham and carried out a series of tests comparing the two. The results surprised everyone. Not only was the carriage with pneumatic tires much quieter and more comfortable, but it was also easier for a horse to pull—much easier.

How was this possible?

It was because of what happened when a wheel rolled over a small stone or bump. A steel wheel had to rise over the stone, then come down on the other side—so pulling the carriage forward meant lifting it slightly. But when a wheel had a rubber tire, the rubber deformed as it went over the stone, while the wheel kept moving forward—no lifting required. Depending on the road, rubber tires could mean as much as a sixty percent saving in energy—a huge increase in efficiency.

Thomson was excited about his discovery. His goal was to put his inflatable tires on steam carriages, allowing them to travel on regular roads more safely, quietly, and at higher speeds.

Thomson had come up with the right invention at the wrong time. Although the manufacturers of the steam carriages would have welcomed Thomson's invention, they were facing huge legal and financial obstacles, challenges that would eventually drive them out of business.

Thomson also considered selling the tires for use on regular horse-drawn carriages. Carriage owners were certainly impressed by the improved ride from pneumatic tires, but his tires were difficult and expensive to make. The price was too high for the average customer, so Thomson eventually gave up on the idea.

Forty years later, in 1887, another Scot, James Dunlop, independently reinvented the pneumatic tire. By then, technology had advanced, new forms of rubber were stronger and easier to work with, and the tires could be cheaply manufactured. Their first commercial success was on a much smaller vehicle—the bicycle.

HOW MANY MILES TO THE GALLOP?

British inventor P. A. Barnes wanted to put horses back on the roads. In 1981, he patented a bizarre horse-powered minibus. Rather than the horse pulling from the front, it rides inside the vehicle beside the passengers, shielded from the elements and providing power to the vehicle by walking on a treadmill belt.

The vehicle has the usual steering wheel and brakes. There is no gas pedal, of course, but the driver can exercise control over the speed of the horse by adjusting a handle connected to a "signal mop" positioned near the horse. "By moving the handle skillfully, the driver can impart signals to the horse." In other words, if you want the horse to go faster, you whack it with a mop.

The driver is presented with a dashboard equipped with a full set of modern instruments and gauges. They show the time, speed, distance covered, something called "collar-push," and horse temperature.

Containers near the rear of the vehicle will collect any droppings, although the current level of droppings is not shown on the dashboard.

The vehicle holds six passengers—everyone has a window on one side, and a horse's sweaty body on the other. They are shielded from the horse (and each other) by a "kick-proof" barrier.

Barnes's patent explains two major advantages to his vehicle compared to a conventional horse-drawn vehicle. One is that, instead of being exposed rough road surfaces and bad weather, the horse can remain in a warm location, walking on a smooth surface. Furthermore, Barnes explains, "even if it is nervous in traffic, it cannot bolt or shy"—or, indeed, move at all from its metal prison.

Another major advantage claimed by the inventor is speed variation—the horse's treadmill is connected to gears controlled by the driver. Considering the large size of this vehicle and the cramped conditions for the horse, we suspect the speeds will range from "slow" to "refusing to move at all."

Enjoy your one horsepower car.

ALL DOWNHILL FROM HERE

James Stack Lauder, known professionally as James Lafayette, was a very successful Victorian and Edwardian photographer. Literally—because his subjects included Queen Victoria and King Edward VII.

With royal patronage backing him up, Lauder became one of the hottest photographers of his day. Aristocrats, theatre stars, and foreign dignitaries lined up to have him take their portraits. He had an artistic eye and was very good at producing that classic, Victorian style of posed photograph.

But despite his huge success as a photographer, Lauder had a secret ambition that most people weren't aware of. He wanted to be an inventor.

Given his line of work, you might expect him to patent something like a new camera, or flash-powder, or a theatrical backdrop. Instead, he was interested in a very different Victorian technology—the bicycle.

Anyone who's ridden a bicycle will have had the experience of riding to the bottom of one hill, then using the boost in speed to go up the next one. Or there's the bad version—riding down the first hill and having to stop at the bottom, so you need to sweat and strain to climb the next hill.

Lauder's idea, in his patent from 1899, was to store the energy used to brake a speeding bike as it descends a hill, then use that stored energy to go up the other side. Today, we'd call this regenerative braking. On many electric vehicles, including e-bikes, when you apply the brakes, a generator turns the vehicle's movement into electricity, which is stored in a battery.

But Lauder's bicycle wasn't electric. He intended to do the same thing mechanically. His idea was to use the frame of the bicycle as a place to store compressed air. Air hoses would be attached to a pair of pistons attached to cranks on the front wheel. If you were going downhill and had to stop at the bottom, you could flip a lever, and the pistons would pump away, reducing your speed and sending compressed air into the frame. When you wanted to start moving again and needed an extra boost to go uphill, you could reverse the system, using the compressed air to power the front wheel.

If the stored air was not needed for climbing difficult hills, Lauder suggested it could also be used to sound a whistle or inflate the tires.

In recent years, a British scientist and inventions expert Adam Hart-Davis examined Lauder's 1899 patent. A keen cyclist himself, he wasn't sure if Lauder's idea was practical but still declared the concept "brilliant."

There's no sign that the bicycle was ever manufactured, but it's clear Lauder put a lot of thought into his patent. While he was photographing the rich and famous people of the day, his mind may have been elsewhere, thinking about speeding away on his air-powered bicycle.

THE FASTEST BIKE IN THE UNIVERSE

Most bicycles use gears—there's usually a big cog next to the pedals and a much smaller one attached to the wheel. When you turn the pedals once, the wheels might turn three, or four, or five times. You can pedal slowly, while your bike moves fast.

It's possible to join several gears together, so each gear spins around at five times the speed of the one driving it. If the first one goes around once, the next goes around five times, and the third goes around twenty-five times.

One inventor decided to take this idea to the limit and built the ultimate fast bicycle. His vehicle consists of eleven sprockets connected to each other by bicycle chains. The gears at each stage share an axle, so each rotation of a small gear with eleven teeth also turns a big gear with fifty-three teeth. Each big gear is connected by a chain to the next small gear in the series.

The speed of the gears is multiplied at each stage. A single turn of the pedals turns the first gear 4.8 times. The next gear turns 23.23 times. The next gear turns 111.98 times . . . the numbers keep increasing, until the final gear, attached to the wheel, which rotates more than 157 million times.

If the cyclist can keep pedalling at a normal cyclist's pace of about 55 rpm—pedalling less than once a second—the bicycle will travel 329,487 kilometres a second. Actually, it will never quite get to that speed because that's faster than light, and that would violate Einstein's theory of special relativity.

That's why the name of the bicycle was the "Relativity Special."

The bicycle was built. As you can probably guess, its actual speed fell short of the speed of light—by a very long way. With all those gears and the accumulated friction from that 157-million-to-one gear ratio, even the best cyclist would find it impossible to move the pedals at all. Even if you had an engine powerful enough to push the pedals, the parts would snap before the rear wheel moved a millimetre.

But the inventor knew all that before he started. He's an artist and sculptor named Kevin Ptak, and he created the Relativity Special as part of an exhibition called Circle of Time. As an artist, Ptak often spent time creating items that weren't useful in the everyday sense, so he decided to take the idea of "realizing useful potential" to its limit, building a bicycle whose mechanism should theoretically move it forward at about a billion kilometres an hour, but that actually can't move at all.

Still, it's nice to imagine. What would happen if, somehow, the bike frame and cyclist were actually strong enough to get it moving? Ptak points out that the rider would still face other catastrophic consequences. Take tires, for example. A set of good bicycle tires costs $100 (US) and lasts about 4800 kilometres. Taking a trip with a single turn of the pedals, you'd have to replace the tires sixty-seven times, at a cost of $3,350. Cycling for a minute would cost you $184,250—a financial catastrophe.

Then there's the matter of friction from the air. Meteors and returning spaceships enter the earth's atmosphere so fast that they're engulfed in a blaze of fire. But their speed is nothing compared to this bicycle. If the bicycle and rider were to avoid being vaporized, they would need what Ptak calls "the heat shield to end all heat shields."

On paper, his bicycle would go as fast as it is possible to go. In reality, Ptak's playful work of art can't move at all. That's the whole point.

But as we'll see in the next chapter, he's not the only inventor to come up with an idea that works in theory but not in practice. And most were not so self-aware.

CHAPTER 9

IT DOESN'T WORK

ONE of the rules of patent offices all over the world is that the idea being patented must have some practical use. You can't patent something that wouldn't actually work. That's why you can't patent a perpetual motion machine—such a machine is impossible.

But patent examiners can't be experts on everything. Some ideas get patented even though they don't work. They are often presented with confidence, and they sound plausible enough to get past the experts.

Here are a few inventions that work better in theory than in practice.

WHATEVER FLOATS YOUR BOAT

In the early 1830s, a young labourer and amateur inventor named Abraham was taking cargo from Illinois to New Orleans.

Along the way the boat got stuck on a small dam, and the lad had to work unloading cargo and making other changes to get the boat free and carry it over the dam.

Riverboats often got stuck back then, and Abraham ran into more boat problems a few years later when, once again, the vessel he was travelling on became stuck. This time, the boat was jammed on a muddy shoal. The captain ordered everyone to put barrels, boxes, and planks under the boat to lift it over the obstacle. They eventually succeeded in getting the boat moving again, but it took a great deal of work.

Abraham eventually quit working as a labourer and became a patent lawyer, but he remembered his experiences on the river and came up with a solution that he turned into a patent of his own.

His idea was to attach a set of flat bellows to the side of a ship. When the ship was sailing normally, the bellows looked like a stack of planks running along the side of the ship, but if the underside of the ship became stuck on mud or rocks, the bellows could be inflated, revealing a fabric balloon between each set of slats. The balloon would expand downwards into the water, forming a set of water wings on each side of the boat. The extra buoyancy from these floats would lift the vessel over any obstructions.

Did these ship-balloons make the inventor famous? Sadly, no. Although the device was never put on a ship and used for real, experts say it has a fundamental flaw: in pumping up the floats, workers would be expending the same effort it takes to lift a fully loaded boat. The time and effort required to fill the floats would have been enormous. It would be easier to do what the riverboat captains usually did—empty the boat of cargo, then shift the vessel with levers.

While Abraham didn't achieve any fame as an inventor, he did manage it in another way, as the sixteenth president of the United States. Abraham Lincoln remains the only American president to hold a patent for his own invention—even if it's a patent that wouldn't work.

CLOUDY WITH A CHANCE OF BLOODSUCKERS

George Merryweather was a British doctor in Whitby, in the north of England. As a medical man in the 1800s, he worked with leeches every day, and he became interested in their behaviours, whether they were sucking the "surplus" blood from his patients, or just swimming around in their jars.

He noticed that the activity of his leeches sometimes changed with the weather. When storms were approaching, it seemed to Merryweather that the leeches became restless, squirming in a distinctive way.

Merryweather decided to harness the sensitivity of the leeches in a new way—as weather forecasters. They would detect subtle changes in the air pressure and provide useful information to their human masters.

As it happens, there was already a device that could detect changes in air pressure and predict upcoming weather. It was called a barometer, and many Victorian homes and businesses had one. That didn't discourage Merryweather, who believed his leeches were detecting something more subtle than mere changes in air pressure—perhaps some sort of electricity in the air.

Merryweather designed a bottle to hold the weather-forecasting leech. If the leech was feeling anxious about the weather, it would try to leave its pool of water and climb to the neck of the bottle, where a mechanism triggered a bell.

The device worked—a little. Unfortunately, Merryweather's leeches were as moody and unreliable as they were sensitive. Sometimes a leech sounded the bell when there was no storm coming. At other times, it might fail to sound the bell when the weather was turning nasty. Merryweather realized he couldn't rely on the capricious whims of a single leech—his machine must use an entire jury—twelve good leeches and true—to make their judgment about upcoming storms and tempests.

Merryweather built a "Tempest Prognosticator," an elaborate device resembling a leech carousel, containing twelve squiggling leeches each in their own containers. He placed each leech in a clear bottle, arranged in a circle so they would be close to other leeches and wouldn't feel too lonely—although that probably didn't make much difference to the leeches, as the eyesight of a medicinal leech is very poor. The movements of each leech could sound a single bell. The more bells sounded, the greater the danger, and if all the bells were sounded, Merryweather said it was a sure sign of an approaching tempest.

Perhaps the leeches were responding to something in the air, but whatever changed their behaviour was not a good indicator of the weather. The device never worked well. However, the Tempest Prognosticator was a gruesome crowd pleaser when it was displayed in the Great Exhibition of 1851. Six million people wandered the halls, and many were drawn to Merryweather's device—although they weren't so fascinated that they wanted to buy one.

WEATHER IN A TUBE

Merryweather's leech-powered Tempest Prognosticator attracted interest from the government of the day, but it faced competition from a different kind of device.

A British admiral named Robert FitzRoy had the job of predicting the country's weather. Big storms sank big ships, and a few hours' warning of an imminent storm could make all the difference.

Working in meteorology, FitzRoy had found that measuring wind direction and using regular barometers could give good predictions of the weather, but although a barometer can give warning of some kinds of stormy weather, it doesn't work in every situation. It is better at indicating the approach of large storm systems than more localized thunderstorms.

But another device seemed to show promise. It was known as a storm glass—a tube filled with water, alcohol, and a few chemicals, which formed white crystals at the bottom. By observing the state of the tube, you could predict the weather: if the liquid went from clear to cloudy, rain was coming. Cloudy with "small stars" indicated thunderstorms. Various other combinations predicted other types of weather.

FitzRoy didn't claim to be an inventor, but he had used the device, trusted it, and believed it was responding to an electrical tension in the air that other devices couldn't detect.

Government officials tested the storm glass, comparing it to Merryweather's leech-based system. They concluded that the storm glass was a more dependable and scientific approach than the leeches. Not only was the storm glass cheaper and simpler to use, but the working parts didn't require a meal of blood to stay alive.

After a series of bad storms caused havoc in 1859, storm glasses were distributed to fishing communities, so crews could check for danger before setting out.

The tube of chemicals might have looked scientific, but it didn't actually work any better than the leeches. Later, when scientists examined the device in more detail, they concluded that the mysterious transformations of the liquid and its crystals were nothing to do with the "electrical tension" FitzRoy had imagined, nor with air pressure. The changes were caused entirely by shifts in temperature. The "storm glass" was nothing more than a low-quality thermometer.

A LITTLE SUBTERFUGE

Some inventions are so far ahead of their time that they can only be the creation of a true genius—unless they're the creation of a fraud. Sometimes it's hard to tell the two apart.

Josef Papp was a Hungarian who emigrated to Canada. In August 1966, he announced that he had invented a new submarine, which he had built in a friend's garage. He showed it to reporters and talked about it on Canadian TV—he claimed his submarine could move with unbelievable speed, fast enough to cross the Atlantic in a day. He even wrote a book on the subject, titled *The Fastest Submarine.*

Papp's wife then called police to say that her husband was missing. So was his submarine. A few days later, he turned up in the Atlantic, not far from the coast of France, floating on a rubber life raft. He explained that he had used his submarine to cross the Atlantic in just thirteen hours. Unfortunately, the submarine didn't make it and had sunk beneath the waves, but the feisty inventor had managed to escape the sinking vessel and survived to tell the tale.

After the submarine disaster, Papp didn't attempt to build a second submarine. Instead, he got money from a Hungarian investor for a completely new project—it was a car engine, powered by fusion. The fuel was sealed inside and would last for a full year, after which it could be cheaply replaced.

Papp and his investor gave a demonstration of the new engine—which looked very much like an old Volvo engine, but Papp said he had made secret modifications. The physicist Richard Feynman attended the event. He was convinced from the start that the device was a hoax—the idea of producing fusion in the way Papp had described made no sense. Feynman challenged Papp, saying the engine was being run electrically from a cable plugged into the wall. Papp pulled the cable to "prove" this wasn't so. Feynman then guessed that the motor also had a battery, so it could run independently. After a scuffle about whether to plug it in again, events turned dramatic. The engine exploded, seriously injuring several people present. (According to some accounts, one person was killed.)

After the disastrous demonstration, lawsuits flew back and forth. Papp blamed Feynman's interference for the explosion. Feynman's university settled with the inventor out of court. Papp and his investor parted ways—the investor had sunk more than half a million dollars into Papp's ideas but didn't see any return from it.

Even today, some people believe Papp "was onto something." Skeptics suggest that he deliberately sabotaged the engine to delay a closer examination. He may have created an explosion that was bigger than he had intended.

Was he a genius or a fake? To this day, nobody is sure how his engine worked, although people have imagined various ways the trick might have been pulled off.

But in the case of his "fastest submarine," the evidence is more damning. On the night Papp was supposed to have disappeared, a man who looked just like him, and carried his passport, had flown from Montreal to Paris, then took a train from Paris to Brest. The train ticket was still in Papp's pocket when he was pulled from his rubber raft. There's no word on the location of his high-speed submarine, but we wonder if anyone thought to check Papp's garage.

BOTTLE OR BURIAL?

Coffins have unpleasant associations. The reason is that when people look at a body in a coffin, they can't help thinking about the way the body and the coffin are going to decay while underground.

This, in any case, was the theory offered by Angelo Raffaele Leero, an Italian living in the United States in 1910. As he saw it, the distress was not caused by grief at the loss of a beloved friend or family member. No, it was just the thought of their future decomposition.

Having identified what he saw as the problem, Leero offered a bizarre solution. No more would dead bodies be placed underground in wooden coffins. Instead, the dear departed would become the "dear remaining": the body would be placed in a huge, glass bell jar, hermetically sealed, with the air replaced by an oxygen-free "preservative atmosphere." Leero believed this treatment would halt decay.

His concept for the glass display case included a wooden stool for the dead person to sit on, including metal rods to hold their head and arms in a natural and lifelike position.

One gruesome detail in his patent was a hole in the centre of the stool. Directly below, in the base of the jar, was a bowl-shaped cavity, intended to catch "the liquid matter usually evacuated from a human body after death."

In other words, to avoid the disturbing idea of a corpse decomposing underground, the inventor offers the horrific alternative of a corpse decomposing slowly in full view of friends and family.

In any case, his ideas about preserving bodies are quite wrong. Many microbes that cause decomposition don't need oxygen—in fact, they don't like it—and Leero's "preservative atmosphere" wouldn't work to preserve the body.

It's definitely an idea that deserved to be buried.

CLEARLY DEAD

Putting a dead body in a big jar is bad enough, but another inventor took the wrong-headed idea much further.

Joseph Karwowski was a Russian living in New York State in 1903. He thought the best way to preserve a body was to embed it in solid glass, like the world's biggest paperweight. Once removed from the destructive effects of air, he believed that the body could be "maintained for an indefinite period in a perfect and lifelike condition, so that it will be prevented from decay."

He had the process worked out. First, he would coat the body in a layer of clear "water glass," a glassy material that flows as a thick liquid. Next, red-hot molten glass would be poured over the corpse to form a large rectangular block. Once the block had cooled off, Karwowski believed the crystal slab would eternally display a perfect human body, and definitely not a gruesome, parboiled, melty-skinned horror.

The illustrations in the inventor's patent show a block big enough to fully enclose a standing adult. It's around 2 cubic metres in volume. Karwowski clearly had no concept of how difficult it would be to produce such a massive block of perfectly clear glass. To put it in perspective, in 2017, the German speciality glass company Schott used all its skills and modern methods to produce a piece of glass with a slightly smaller volume for use as a telescope mirror. It took a full year to cool it down without cracking.

Not only would manufacturing the crystal corpse block be a nightmare, so would moving it—the glass monolith would weigh somewhere in the region of 5 tons. A paperweight like that will not only flatten your documents, but also the desk they're sitting on.

Obviously, this project is only for the super rich. But for those on a tight budget, Karwowski also offered a smaller version. A grieving family could be consoled and comforted by gazing at their relative's severed head.

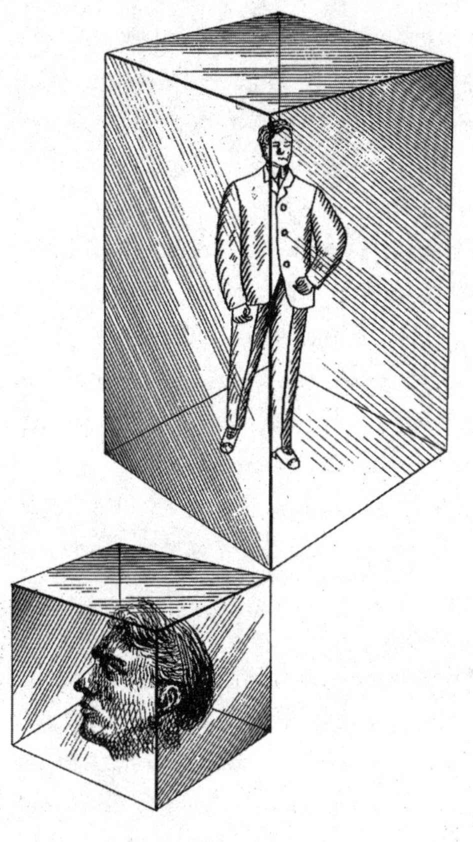

THE WORLD'S HARDEST PUZZLE

As a general rule, an invention that doesn't work won't do well in the marketplace—but there are exceptions, and one of them came from the mind of the legendary inventor and game designer Sam Loyd.

Everyone has seen the "Fifteen Puzzle," also called a sliding block puzzle. In the modern plastic version, fifteen plastic tiles are arranged in a frame, making a four-by-four grid with one empty space. The puzzle solver slides blocks around to get the tiles in the right order.

This puzzle was invented sometime in the 1870s by a New York postmaster named Noyes Chapman. His version used wooden blocks rather than tiles. The puzzle quickly became a public obsession, just like the Rubik's Cube a century later. The craze spread like wildfire across America and Europe—newspaper editorials complained that people were sliding puzzle blocks when they were supposed to be working.

In 1879, a couple of mathematicians named Johnson and Story looked at the new fad and wrote a paper about it in the *American Journal of Mathematics*. They were curious to see if the numbers could be arranged into any order or if there were some impossible combinations. It turned out that any starting arrangement of blocks belonged to one of two sets, which they called "positive" and "negative." With enough slides, a "positive" arrangement could be turned into any other "positive" arrangement, and a "negative" arrangement could be turned into any other "negative" arrangement. But if you arranged the blocks in a "positive" arrangement, no amount of sliding could get them into a "negative" arrangement, and vice versa—it was mathematically impossible.

Puzzle maker Sam Loyd was a mathematician as well as a huckster, and he may have read this article. He thought of a clever way to cash in on the block-sliding craze.

Loyd produced a version of the puzzle with the numbers laid out from one to fifteen, but with the numbers fourteen and fifteen reversed. Instead of the bottom row of the puzzle showing thirteen-fourteen-fifteen, it showed thirteen-fifteen-fourteen. He challenged solvers to rearrange the puzzle so all the pieces were in the right order. To sweeten the deal, he offered a large cash prize to the first person who solved the puzzle and could show how they did it.

Of course, Loyd knew he was at no risk of losing his money. By reversing the final numbers, he had made his puzzle unsolvable, but he also knew you'd need to be a mathematician to understand why.

(The same trick will work on a Rubik's Cube. Removing one piece and putting it back with the colours reversed can produce an unsolvable puzzle.)

Loyd's version gave the craze a new lease on life. People bought Loyd's new version of the puzzle, sliding the little tiles and trying to find the money-making solution. Some said they had solved it but couldn't show how. Many others came tantalizingly close—they'd get fourteen and fifteen into the right places, along with most of the other numbers, but there was always one infuriating reversal elsewhere in the grid.

Loyd tried to patent his version of the puzzle. The regulations at that time required applicants to submit a working model, so the patent official asked Sam Loyd to provide one.

Sam Loyd said, "Working model? No. It's mathematically impossible. The whole point is that it doesn't work."

The patent clerk considered the matter and said, "If it doesn't work, it can't be patented."

"Fair enough," said Loyd.

CHAPTER 10

GUYS WITH WINGS

DAEDALUS was the flying inventor from Greek mythology, who has remained an icon for inventors ever since.

The oldest stories about him go back nearly 3,500 years, to the Mediterranean island of Crete, and he was a legendary figure even then, although in most of the earliest stories, he's associated more with woodwork than wings. He is supposed to have invented most of the tools used by carpenters, including the axe, the saw, and glue. And you know those sails on a ship's masts? Also his idea.

It was said he could carve impressively lifelike statues. His name is connected to a Greek festival, where oak trees (widely considered to have magical powers) were cut down and carved into wooden statues called "daedala."

It's quite possible the first stories are fuzzy memories of some real inventor, an expert carpenter, but if that's true, he lived a very long time ago. As the centuries rolled by, people told new stories about him. He turned into a sort of MacGyver figure, getting into scrapes and inventing his way out of them.

Of course, the most famous of these stories is about how Daedalus built a set of wings. It happened when Daedalus and his son Icarus were being kept prisoner in a tower (or, in some versions, a maze). Escape seemed impossible until Daedalus used his inventive powers—he built a pair of wings for himself and another pair for Icarus, and he carefully attached feathers to them. Strangely, despite being the legendary inventor of glue, he didn't bother using it in this case, sticking the feathers on with beeswax instead.

After warning his son not to fly too high, the pair took off and flew away. The foolish Icarus ignored the father's advice and flew so high that the hot sun melted the wax. The feathers fell from his wings, and the boy fell from the sky.

Despite the limitations of the wings, they worked well for the father, and he flapped his way to safety. The moral of the story is that you must control your ambitions and "not fly too close to the sun."

But it also seems to carry the message that, if you control your altitude, human-powered flight can be safe and effective. At least, that seems to have been the takeaway for generations of would-be flyers.

BLADUD, KING OF THE AIR

A less well-known early inventor and flyer from myths was King Bladud. He was a legendary king of the Britons. According to medieval historians (who were generally not too reliable) he lived around 800 BC.

Bladud came up with two inventions. His first was a cure for leprosy. As a prince, he had contracted the disease, and he ended up having to leave the comforts of the palace and work as a swineherd. One day, he noticed that pigs rolling in a certain patch of hot mud came away with improved skin. He tried it himself, and the treatment cured his leprosy. He later became king, and founded the city of Bath, to share the cure with other sufferers.

His other invention involved flying. Bladud was a keen practitioner of necromancy. During his conversations with the spirits of the dead, they revealed to him the secrets of flight. He followed their instructions and built a pair of feathered wings, then leapt from a great height for a flight across London.

Perhaps the spirits had just been having fun with Bladud, though, because unlike Daedalus, his wings didn't work at all. He came tumbling to the ground, breaking every bone in his body. Bladud was Bla-dead. His kingdom then passed to his better-known son, King Lear.

SPANISH FLY

Around the year 800, back when Spain was a Muslim country, a man named Abbas ibn Firnas was the country's Thomas Edison. He invented a clock and a metronome. He also ground his own lenses and built astronomical instruments.

Like many other inventors throughout history, he decided to give flying a try, and he took the usual bird-inspired approach. He built himself a set of wings covered in vulture feathers, then found a tall place to jump from.

Accounts of the flight are garbled—some say he jumped off a minaret in Cordoba, others that he leapt from a mountain in Yemen.

Wherever it happened, the accounts agree that the feathered wings worked for this Spanish Daedalus. The inventor had followed his dream of flight and had managed to fly. Where he fell short was in following his dreams of landing softly. He came down hard. He didn't die, but he suffered a serious back injury. Commentators at the time said that he would have landed better if he had strapped on a tail.

Still, many people believe the event really happened, and it sounds plausible that Abbas ibn Firnas made a successful flight followed by a not-so-successful landing.

Baghdad International Airport has a statue of the inventor preparing to take flight. Because when you're a nervous passenger about to board an aircraft, you want to be inspired by the image of a winged man who is moments from crashing.

THE FLYING MONK

A couple of centuries after Abbas ibn Firnas had made his abortive flight, a young British monk tried something similar. Eilmer was a monk at Malmesbury Abbey, in the southwest of England. He had heard the story of Daedalus building wings, and he was inspired to try it for himself.

Around the year 1000, Eilmer built a set of wings and climbed to the tower of Malmesbury Abbey to try them out. According to an account by a medieval historian known as William of Malmesbury, Eilmer jumped off the tower and glided about a furlong from the tower. (If you don't know how long a furlong is, it's about 377 cubits, or 7920 inches, or 200 metres.)

Just like Abbas ibn Firnas, Eilmer had a good flight followed by a bad landing. He hit the ground hard, breaking both his legs.

Eilmer survived, but he was "lame ever after." Despite his injuries, he lived to a ripe old age and shared his stories with the other monks—some of whom may have shared them with William of Malmesbury, a century after the event. The account is brief but plausible.

Apparently, Eilmer later said he would have landed better if he'd added a tail—the same improvement that critics of Abbas ibn Firnas had suggested. Everyone back then seemed to think birds landed on their tails.

Eilmer the flying monk is considered Britain's first aviator.

F FOR FLY, FLAP, AND FALL

A very brainy scholar named Abu Nasr al-Jawhari lived around a thousand years ago in what is now Kazakhstan. He was an expert in the Arabic language, and it seems he was dissatisfied with the way other people wrote Arabic because he spent years writing a huge book explaining how people should write it. The title of his book translates as *The Crown of Language and the Correct Arabic* and it would become the most important Arabic dictionary.

This was a monster work for the time. Al-Jawhari wrote entries for 40,000 Arabic words. But how do you organize so many words? There wasn't a standard way of doing it back then. Today, we're used to sorting words alphabetically, but that was a later choice. In the ancient world, the dictionaries had often been unsorted lists, or the words were sorted in categories or by the words they rhymed with.

Al-Jawhari's big idea was to sort the words alphabetically, by the main Arabic "root." This makes more sense in Arabic than English, but it would be like having an English dictionary where "overfly," "flight," and "dragonfly" were all grouped together under F for "fly."

He got most of the way through his dictionary. One of his students completed the work. The book was used for centuries, and his sorting system is still used by Arabic dictionaries today.

The reason Al-Jawhari didn't complete the work himself was that he suddenly had the urge to take his studies in a completely different direction.

Maybe sitting in a library sorting thousands of words in *Correct Arabic* started to take a toll on Al-Jawhari, or perhaps he had just spent too long indoors in a library, but whatever the reason, a few years after 1000 AD, he suddenly had the urge to fly. He built himself a set of wings and climbed to the roof of his local mosque, then took a short flight—in a mostly downward direction. He was killed when he hit the ground.

What's the Arabic spelling for "Ouch"?

RENAISSANCE JAILBIRD

Da Vinci was famous for his sketches of invention ideas. One shows a man strapped to a machine with large batlike wings, called an ornithopter. His inventions were not widely known in his day because da Vinci kept his journals secret—but he was not the only person thinking of flight this way, and another artist from around the same time and place was quite open in describing his plans.

Like da Vinci, the goldsmith Benvenuto Cellini was a Renaissance man who combined a love of art with an interest in practical engineering problems. He is famous for his egotistical autobiography, which describes his own wonderful artworks and his brave adventures.

Cellini claimed that he was once wrongfully imprisoned by the pope.

The official in charge of his prison, known as a castellan, enjoyed the goldsmith's company and spent hours talking with him. Unfortunately, the castellan suffered from periods of madness where he would imagine his human body had taken on a different form. His servants saw signs that another bout of madness was approaching and asked the prisoner Cellini to talk to him and keep him occupied.

In one conversation, the castellan asked whether Cellini had ever thought about flying. Cellini said yes, he had thought about it, and although most people considered human flight to be impossible, Cellini explained he had always been able to accomplish things that were beyond the powers of lesser men, and he was sure he had the mental skills to invent a way of flying, not to mention the physical skills and courage to put it into practice. (Cellini was never shy when describing his own virtues.)

The castellan was fascinated. He asked Cellini to explain how he would go about it.

Cellini replied that he would use his skills to imitate nature. In his view, the most perfect flying creature was the bat. He would use waxed linen to create a pair of batlike wings, and once they were built, he was quite sure that he could fly from the prison to the nearby neighbourhood of Prati—a distance of about a kilometre.

Cellini's ideas sound like da Vinci's—the idea of flying like a bat rather than a bird seems to have been in vogue at that time. However, as it turned out, Cellini might have been wiser to do what da Vinci did and keep his ideas secret. As soon as he started talking about bats, the castellan became very excited by the idea, and from then on, he believed he himself was a bat.

The castellan told Cellini that he would keep him tightly locked up so he couldn't fly away. Cellini said that if that's how he was going to be treated, he would escape. "No, you won't," said the castellan. "If you fly away, I'll come flying after you."

Cellini was kept in his cell and was now being closely watched. He set to work on his escape plans. He somehow managed to collect the items he needed. He secretly stole a pair of pincers from one of the guards who did carpentry work. He also collected a quantity of wax and carefully cut up the cloth from his prison sheets.

It would have been great if Cellini had tried to fly away on wings, but he wasn't that dumb. Instead, he used the pincers to pull out the nails on his window bars and used the wax to make fake nail heads so the change to the bars wouldn't be noticed. He then made a long rope from the cloth strips, so he could descend from the tall tower.

After Cellini escaped, he claimed the delusional castellan wanted to pursue him by jumping from the tower and flying after him, and the castellan's servants had to fight with him to stop him from attempting a bold human flight experiment using only his flapping arms.

Whether Cellini's account is true or not, he certainly wrote it in the 1500s, and it gives an insight into the way people thought about human flight in those days. His flight of fancy may not have produced any real wings, but his efforts did allow him to "fly the coop."

THAT'S CHEATING!

In 1889, Reuben Jasper Spalding had an idea that would allow him to achieve independent flight. It consisted of two large, feathered wings, and a long, feathered tail, all attached to a harness. By moving his arms and legs, the flyer could flap the wings.

Most of these kinds of inventions were disasters. The wings were usually too small or too flimsy to support the weight of a human. But Spalding knew about the problems and accounted for them. He made his flying kit airworthy by giving the flimsy wings a little outside help. His idea was to attach the whole contraption to a large helium balloon.

Of course, if you're hanging from a balloon, you don't really need any wings at all—you could fly just as well by carrying two feather dusters or just flapping your hands. In fact, since most of your movement will be determined by the wind blowing the balloon, you might as well just dangle limply and go where the balloon takes you, but where's the drama in that?

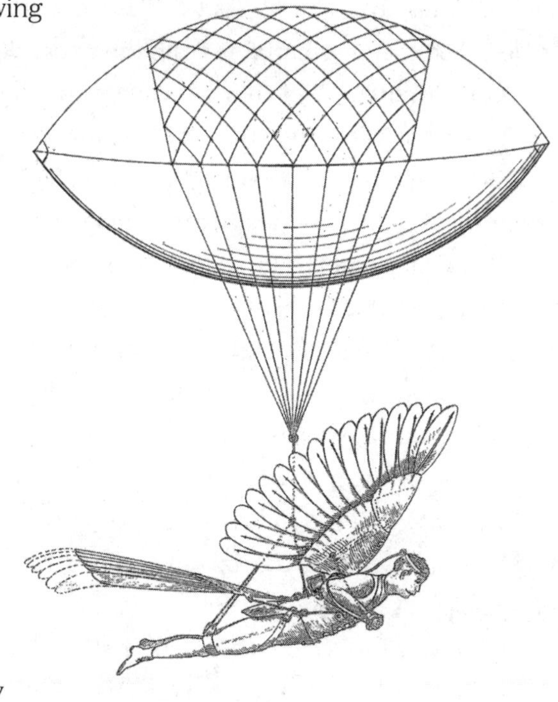

There's no record of Spalding ever flying his machine, although he seems to have had many other adventures: he spent some years in the US military and later made money in the California Gold Rush of 1849. Although he remained fascinated by the idea of true human flight, he didn't get to see it. He died in 1902, a year before the first flight by the Wright brothers.

GERMANY'S BATMAN

Most of the time, if a person straps wings to their arms and tries to fly, the results are going to be messy. But in the 1890s, a German inventor named Otto Lilienthal took a more careful approach. He had a background in engineering, and he designed a set of wings based on his study of bird wing aerodynamics.

Many of the other bird wannabes we've looked at chose the tallest cliffs and towers for their flights, but Lilienthal started sensibly with baby flaps, jumping off a hill on the side of a sand pit. His homemade wings looked strange and batlike but worked in a similar way to a modern hang glider, with the pilot hanging from the underside and shifting his weight to steer. He soon got the hang of things and was actually flying.

Lilienthal was a big, athletic Prussian. Although most people haven't heard of him today, he made thousands of flights in the 1890s, and his pictures were in all the newspapers and magazines. He became world famous as "The Flying Man." He wrote about the scientific principles of flight, and even manufactured copies of his set of wings, so others could fly. The wings needed a catchy name. He came up with *Lilienthal Normalsegelapparat*. Evidently, he was better at engineering than marketing.

After all his success gliding on wings, Lilienthal started thinking about the next steps—powered flight. He had some ideas. In addition to his expertise in aerodynamics, he also happened to be an expert in steam engines, so it's not hard to imagine where things might have gone next. A steam biplane, anyone?

However, it was not to be. Lilienthal had made unpowered flight seem like a normal thing, but it was still a dangerous hobby. One summer day in 1896, he was enjoying a routine flight using his famous wings when he suddenly pitched upwards, and his wings stopped providing any lift. He started to drop. He tried to correct the problem, but he was unsuccessful and fell to the ground. He had not been flying at a great height—only 15 metres or so—but it was high enough. He fractured his neck in the crash. He was taken to hospital and died the following day.

His death had a chilling effect on German aviation—it seemed to prove to the public that flying was too dangerous for anyone to try. After all, if even the great Lilienthal couldn't master it, what chance did lesser aviators have? At his widow's request, his remaining flying machines were burned. So were all the papers containing his new designs—we'll never know what he might have come up with next.

But the fatal accident also spurred new efforts. In the US, a bicycle salesman named Wilbur Wright saw the headline about Lilienthal's death. According to him, this was the moment he started thinking seriously about building a flying machine of his own.

CHAPTER 11

LOOKING GOOD

IN the popular imagination, inventors are often associated with complex machinery and weird gadgets.

But inventors are not merely hard-nosed engineers. They can be sensitive, too, and appreciate the wonder and beauty of the human form.

Some have even thought about ways to maintain and enhance a person's appearance . . . using complex machinery and weird gadgets.

MOUSTACHE PROTECTORS

A Victorian man loved his moustache. It was the symbol of his magnificent manliness. But it didn't look so good if his sculpted whiskers got soaked with turtle soup.

Inventors lavished attention on this problem, and there are many inventions to accommodate the mighty moustache, including scores of "moustache guards" and a variety of spoons and drinking vessels with attachments (usually a barrier with a hole in the bottom) to keep the moustache separated from soups, drinks, and the nightmare of sticky desserts.

In 1882, Charles Miller invented a moustache guard—a metal device resembling a rectangular paperclip that held one side of the moustache in place. The wearer needed to attach a clip on each side to get full protection, which must have looked ridiculous.

In 1876, V. A. Gates patented a cloth guard—like a surgical facemask but covering only the area between the mouth and nose. Two cords went over the ears and a cloth strip covered and protected the delicate moustache from spills and airborne viruses.

William Masterman's 1889 patent used a wire frame that went over the ears like a pair of eyeglasses, but instead of riding over the nose, it dipped down beneath it. The moustache was tucked into a set of fine wires running near the frame, keeping the precious whiskers away from the soup—although if the glasses had fallen off and into the soup, they would certainly have dragged the moustache with them.

An invention by a man named J. Dal seems to be nothing more than a ordinary women's hair clip, fastened to a moustache. This hardly seems to deserve a patent, but this was a man's hair, so it was given one anyway, in 1896. Again, one clip went on each side, dangling like Christmas ornaments from the facial-hair tree.

In 1901, Horace Joseph Mier of New Jersey invented a moustache guard for cups, which looks so similar to the plastic lid on a modern coffee cup that we have to wonder if it was the inspiration for it. However, unlike a plastic coffee lid, his cover had no way to seal against the edge of the cup. While the user's moustache would have been well protected, he would need to keep the lid firmly pressed against the cup to stop coffee pouring over his silk bowtie and fancy yellow vest.

DIMPLE MAKER

What could be lovelier than a radiant smile? A radiant smile with dimples, that's what! And it was hard luck for those not blessed with natural dimples. But German inventor Martin Goetze was there to save the day.

As Goetze said, "In order to make the body susceptible to the production of artistic dimples, it is necessary, as has been proved by numerous experiments, that the cellular tissues surrounding the spot where the dimple is to be produced should be made susceptible to its production by means of massage."

So, in 1896, Goetze patented his dimple maker to carry out this massage. It resembled a hand-cranked drill, with a built-in measuring device to ensure that the position of the dimple was placed with German precision. Once aligned, the operator pressed on the drill and turned the crank, and in a few short minutes the subject's cheek would sport a lovely dimple . . . or was it just a blister?

I CAN'T WAIT TO SEE THE BACK OF YOU

If you've ever tried to trim the back of your neck, you'll know how difficult it can be to see yourself from behind. Getting a good look usually involves a series of contortions standing in front of a large mirror, while angling a hand mirror behind you, or vice versa.

Many people dreamed they could wear a mirror on their body, so they could see their own back. At least, inventors hoped it's what many people dreamed, because that's the invention they wanted to sell.

In 1905, Emmie Alice Thayer and Emily Waitee Thayer of Bellows Falls, Vermont, designed a mirror that attached to the front of the body using a pair of hooks over the ears. The user could now stand with her back to a wall mirror and see herself from behind, while keeping both hands free. The design also allows a very vain person to keep the mirror hanging in front of them as they walk around, enjoying the look of their own face.

In 1952, Edward O'Brien invented a more complex rear-view mirror. It strapped on like a corset over the upper body. A row of three round mirrors provided a view of the back of the head, which the wearer could easily see by looking into a regular bathroom mirror. O'Brien's device even included a light. Unfortunately, the body straps are complex, and it looks as if you'd need the help of a mirror or two just to strap yourself into it.

In recent years, people in China have become interested in the health benefits of walking backwards. Whatever the positive effects on health, the negative ones include falling into holes and colliding with solid objects, so the fad has led to a recent flurry of head-mounted mirror inventions. One mirror mounts on the head like a cap; another takes the form of specialized eyeglasses; yet another wraps around the neck.

Even if the backwards-walking craze fades away, this generation of Chinese citizens will have an easy time trimming their neck hair.

CUPID'S BOW LIPS

In the 1920s, "Cupid's bow" lips were all the rage. In case you missed the memo, those are full lips that rise to two peaks below the nose. Mary Pickford had Cupid's bow lips. So (appropriately) did "it girl" Clara Bow. All the most lauded beauties of the day had these lips, and if they didn't, makeup made it look as if they did.

But what if a woman wasn't naturally blessed with this coveted shape of lip? A Kansas inventor named Hazel Mann Montealegre offered a simple solution. In 1924, she patented a metal clamp with a rubber clip. The user fastened the device onto her upper lip. It held the sides of the upper lip in place, but pulled the centre down, squeezing the lips into the correct shape.

Mann claimed the lips would hold the new position for some hours, although it seems likely that attaching a narrow, spring-loaded clamp to the middle of your lip might run a risk of swelling and bruising. Then again, that could yield a different movie star look—a Charlie Chaplin moustache.

AUTOMATIC HAIRCUTTER

There are various ways for a barber to cut a man's hair—scissors, an electric hair trimmer, or even a razor. But all these methods pose problems. They depend on the skill of the barber. They make a mess. The blades get dull. And it's so slow—cutting first one area, then another.

In 1951, inventor John Boax of Pennsylvania invented a device that took a revolutionary new approach to the haircut. It was a helmet resembling the hooded hair dryers used in ladies' hairdressing salons, but instead of blowing a man's hair, it sucked at it, using a vacuum cleaner-style hose.

The individual hairs were drawn through a series of narrow metal vanes inside the helmet. The length of these vanes was carefully designed so the hair on different parts of the head would be drawn to the edge of the helmet at different lengths. Following the men's fashions of the day, the hair was to be long on top and in front, and short at the back and sides. There were no cutting blades. Instead, the inner edge of the helmet contained an electric heater. When the hair touched it, it was burned off.

What was supposed to happen—and what the inventor likely envisioned—was that the device would be carefully adjusted and fitted on the customer's head, then the vacuum and heater power would be turned on, and in a few short seconds, every hair on the customer's head was burned off at the correct length. Instantly, a perfect haircut.

It's not hard to imagine the potential downsides. A machine with the power to singe an entire head of hair all at once would require a huge heating element, so this device would get much hotter than any hair dryer. Essentially, the customer is sitting with his head inside an oven. Even with a vacuum drawing the air away, it might get uncomfortable, and it would get a lot more uncomfortable if his hair gets stuck to the metal or catches fire. This instant haircutter is likely to deliver a "short back and fried."

MIRROR KNIFE

One of life's common embarrassments is discovering you've spent the evening with food stuck on the edge of your lips or between your teeth. But how do you check, when there's not usually any way to look at yourself while dining, and pulling out a hand mirror might look as if you're doing personal grooming in the dining room—an etiquette no-no?

In 1908, Elmer Walter from Pennsylvania turned his mind to this problem. His solution was to attach a mirror to a dinner knife. The mirror would be tiny and discreet, placed at the base of the handle.

With Walter's mirror-handled knives, diners would no longer need to take an unneeded trip to the washroom to check the state of their teeth. Social embarrassment averted!

We wonder, though, how diners might have reacted to a person clutching a knife in their fist as they gaze down at it, while baring their teeth maniacally.

CHAPTER 12

MEDICAL MIRACLES

WITH some inventions, the question of whether it works or not is a life-and-death affair. For example, flying machines can be dangerous to test. The same is true with many medical inventions.

But the medical inventor who thinks up a new medical device or treatment has some advantages over the aeronaut. The flyer usually risks his own life, but for the medical inventor, it's the patient who takes all the risks. Although, as we'll see, that's not always true.

SAFER SURGERY THROUGH TOXIC CHEMICALS

Joseph Lister was a nineteenth-century doctor who had a particular interest in germs. Listeria is named after him.

He was interested in chemicals that might kill germs. He had read that a chemical called carbolic acid was useful for reducing the foul stench of sewers. Lister reasoned that if the smells were coming from harmful bacteria, the carbolic acid might work by killing germs.

He soon had a chance to put his ideas to the test. In 1865, he was working at a hospital in Glasgow. An eleven-year-old boy was brought in for treatment. His leg had been crushed under a cartwheel. It was a compound fracture, meaning that the broken bone was jutting through the skin. In those days, doctors knew that putting the bone back in place would likely lead to gangrene and the rapid death of the patient. It was safer just to amputate the leg—even though a lot of people died that way, too.

But Lister thought that perhaps the leg could be saved if the wound was disinfected. He carefully soaked the area with carbolic acid, then set the bone. He watched carefully for the first signs of infection, but there were none. After a few weeks of recovery, the leg was fully healed, and the boy walked home.

Excited, Lister tried the same treatment on other patients with badly broken limbs. A combination of setting the bone and disinfecting the wound with carbolic acid worked equally well on them.

Lister decided to take things further. After all, if this substance could stop the spread of infection after accidental injuries, it should also stop infections after surgery. With the miraculous carbolic acid at his disposal, he planned to kill all the germs in the operating room.

He gave instructions to the surgical team. First, he said, all the tools must be sprayed with carbolic acid. Bandages should be soaked in it. Surgeons should also wash their hands in it. And because germs might be floating in the air, he invented a steam-powered sprayer that sent a mist of carbolic acid into the air during the operation.

The surgeons went along with Lister's instructions and carried out some operations. It actually worked very well. The number of people dying after operations dropped from forty-five percent to fifteen percent.

A breakthrough in medicine, right?

Well, yes and no.

The trouble is that carbolic acid, also called phenol, is an unpleasant chemical. It is extracted from tar or oil products, and it has a sickly sweet, tarry smell. In large quantities, it is very toxic. When you breathe it, it irritates your lungs. Put it on your skin and it makes the skin turn grey and scaly. Contact with the skin also creates sores—when Lister had tended to the patient with the broken leg, the side effects included chemical burns.

This is the substance the surgeons were slathering on their hands and breathing in the operating room. Sure, the patients might recover, but the surgeons walked out of the room coughing and hacking, their skin a mass of scabs and sores.

Lister's carbolic acid sprayer was definitely a mixed blessing.

As it turned out, spraying the air to keep it germ-free was not necessary. Lister later discovered that most of the bacteria floating in the air didn't cause many infections. They could use the disinfectant on the surgical instruments and the patient's wounds and achieve results that were just as good. His colleagues must have been delighted to learn that the sores, blisters, and coughs they had endured were no more than one of Lister's little errors.

The surgeons abandoned the sprayer—and so did Lister. But the idea of sterilizing surgical instruments and swabbing the patient's skin to kill germs is still with us today.

CARBOLIC SMOKE BALL

During the COVID pandemic, some people unwisely tried inhaling disinfectants to prevent illness. It was not a smart idea: as Lister's experiences showed, chemicals that kill germs can also damage healthy cells.

People had tried a similar remedy more than a century earlier. Around 1890, Europe was reeling from an epidemic known as the "Russian flu." As with COVID, many people were getting sick from this disease, and some died. In fact, some researchers now think it wasn't influenza at all, but a coronavirus like COVID.

Carbolic acid was then widely used as a disinfectant, and a man named Frederick Augustus Roe claimed that, since spraying carbolic acid could kill germs on surfaces outside the body, breathing it into your lungs would be a good way of fighting germs once they were on the inside.

He mixed up a powder made from a few plant compounds and carbolic acid in crystal form, then put them inside a rubber bulb, with a short tube attached. He called his invention the "Carbolic Smoke Ball." The user put the tube into a nostril, squeezed the bulb, and inhaled a cloud of white powder that was supposed to kill any germs trying to make a home in their nose and lungs.

This is a terrible idea. Certainly, carbolic acid kills germs. So do fire and boiling water, but inhaling them is no way to cure a cold, and carbolic acid is not much better. The chemical destroys tissues, and it is toxic if you swallow any. Even in the late 1800s, doctors knew this and were campaigning to have the substance controlled.

But Roe didn't worry about medical naysayers—whatever qualms people might have about his invention were nothing a little advertising couldn't overcome. His business, the Carbolic Smoke Ball Company, ran advertisements in magazines and newspapers,

Roe's company claimed that if you used the bulb regularly, three times a day, you would be protected from coughs, colds, asthma, bronchitis, laryngitis, diphtheria, croup, whooping cough, neuralgia, snoring, sore eyes, and headache. Oh, and influenza. The smoke ball provided protection from the Russian flu.

The print advertisement was covered with testimonials from doctors, ministers, aristocrats, and celebrities. They boasted that it had been supplied to the Duke of Edinburgh, Lord Tennyson, the Lord Chancellor, the Earl of Derby, and many more.

Roe's company stood behind the product. Thousands of the balls had been sold, the advertisement said, and not one person had become sick. In fact, Roe guaranteed that anyone who regularly used one of the smoke balls would be protected from all the diseases listed on their advertisement, and if he was wrong, he would give a reward of £100—the money had been set aside in the bank.

Mrs. Louisa Elizabeth Carlill was not an aristocrat or a celebrity, but she wanted to guard against illness. In November 1891, she bought one of the Carbolic Smoke Balls and used it three times a day, according to the instructions. The process was unpleasant, but people expected that from their medicines. Two months later, she got the flu.

Once she'd recovered, Carlill wrote to Roe's company and told them what had happened. She claimed her £100 reward. The company ignored her letter. She wrote a follow-up letter, and they ignored that, too. She sent a third letter. This time, the company replied, saying that they knew for a fact that the Carbolic Smoke Ball was effective, so they would not be paying out.

Carlill was outraged at their response. As she saw it, Roe and his company had made a promise and reneged on it. Her husband was a solicitor, so she launched a court case against Frederick Roe and his company.

The company's lawyers tried to argue that the claims in the advertisement were "mere puff"—like the clouds of carbolic powder the ball dispensed. It shouldn't be taken seriously, they said. After all, this was an advertisement, not a legal contract.

The judges disagreed. They said that the claims in the advertisement were indeed a contract with the many people who bought the product. The company had promised to pay out if it failed and said they had money set aside in a bank account. It was nonsense to call it "mere puff." It sounded like a serious promise.

Mrs. Carlill had trusted Roe's enthusiastic claims, paid for the smoke ball, and used it as instructed. It had failed to live up to the claims. The company was ordered to pay her compensation.

It was a very important case in contract law, and even today law students are taught about the Carbolic Smoke Ball case.

Although lawyers may have learned a lesson from the case, Roe didn't. The following year, he literally doubled down, running the same advertisement, with the reward now increased to £200. His company even used Mrs. Carlill's case as part of the advertising, saying that thousands of smoke balls had been sold, but only three people had claimed a reward, so it was clearly highly effective.

Mrs. Carlill wasn't the only person let down by the smoke ball. In fact, one of the VIPs on the advertisement, the Earl of Derby, contracted Russian flu and never recovered. His name was Edward Stanley, and after he died, his title passed to his sports-mad younger brother, Frederick Stanley. Frederick ended up becoming governor general of Canada and donated the original Stanley Cup. If the Carbolic Smoke Ball had worked, the most famous award in hockey might never have existed.

But karma has a way of catching up to people. Not long after his smoke ball escapade, Roe died from tuberculosis at the age of fifty-six.

As for Carlill, she lived a very long life, eventually dying at the age of ninety-six from old age . . . and influenza.

FRESH BLOOD

Alexander Bogdanov was a Russian doctor with big ideas that went far beyond medicine. He had been one of the founders of the Bolshevik party and the main rival to Lenin, and wanted more power for the intellectuals in the group.

He was a writer, too. He wrote a science fiction novel, *Red Star,* about a scientist and revolutionary (much like himself) who goes to Mars and discovers a communist paradise.

His medical interests were no less revolutionary. He was obsessed with blood, and he set up Russia's blood transfusion service. This wasn't just to help people in operations or accidents—he believed that a person's vitality came mostly from their blood. If you could set up a nationwide system of blood transfusions, you could share health as well as wealth.

Bogdanov believed that if you could put a young person's blood into an old person, the old person would be rejuvenated. Bogdanov himself was already in his early fifties, so he had some personal interest in this immortality scheme. Like a scientific Dracula, he tried a transfusion himself, taking just a little blood from a younger person and injecting it into himself. He felt fine, so he tried it again, and then again. He was pleased with the results. He was convinced the procedures had improved his eyesight and that his balding had slowed. Others signed up for his age-defying procedure, including Lenin's sister.

One of his students suffered from malaria and tuberculosis. Bogdanov saw a wonderful opportunity here. A transfusion with the student would put the student's young blood into Bogdanov's old veins, rejuvenating the scientist further. And the student might benefit too—Bogdanov believed he had a natural resistance to tuberculosis, so putting his blood into the student would allow the young man to recover from the diseases.

The student was fine—in fact, he recovered from his illness—but Bogdanov's overconfidence caught up with him. A few hours after the transfusion, he became very ill—probably because the blood transfused into him was incompatible with his blood type. Bogdanov rallied long enough to take notes recording his body's reaction, then died.

MOUTH EXERCISER

Some doctor-inventors have fixated on a narrow area of health for their medical inventions, in the hope that addressing a small problem would also fix a bigger one. In 1923, Charles G. Purdy was a physician in his seventies, living in Brooklyn, NY. He was alarmed at the state of modern mouth muscles.

All around him, he saw people with decaying or crooked teeth, small jaws, and gum disease. He saw weak mouths everywhere. He also believed he knew the cause—the modern diet. Food was too soft, so people didn't have to chew their food anymore. Why, the human mouth was fading away!

Of course, he had a solution: an oral exercise regimen. His mouth exerciser consisted of a bite plate, custom designed to fit a person's teeth. It was attached to a cord, which was tied to a spring and could be fastened to a wall. Once it was hitched up, the user would clamp down their teeth on the plate and jerk their head backwards to get a good mouth workout, improving circulation in the area.

He also invented a two-person version. Each person was given a bite plate, connected together by a spring. The picture accompanying the patent shows a man and woman enjoying the device, as they engage in a furious tug-of-war, like Lady and the Tramp if they behaved like real dogs.

There are some obvious hazards with this invention. Using your teeth to pull with great force might run the risk of extracting one or two of them. There's also the danger that the cord could break, sending a steel spring into your face. Then again, those who engage in the two-player game might use this strategy for fun. If they are losing their tug-of-war, they only need to open their mouth to send a saliva-soaked bite plate rocketing at their opponent's mouth.

The Purdy Exercising Device doesn't seem to have been a successful product.

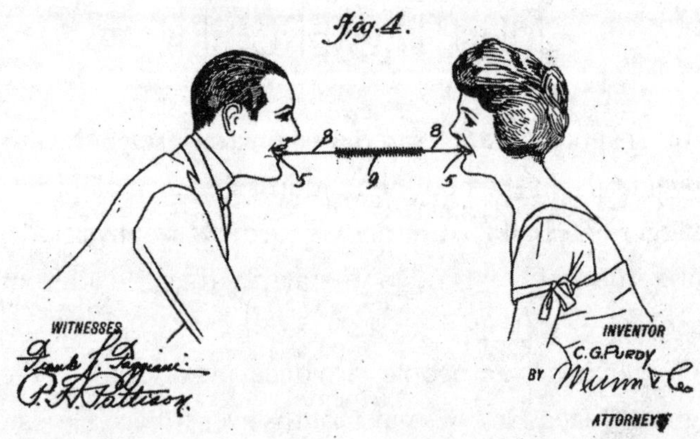

When Purdy wasn't worrying about his fellow citizens' mouth muscles, he was concerned about their driving habits. He campaigned against unsafe drivers. In 1930, he publicized a number of rules for drivers to follow: "Drive like a gentleman and don't show off" was adopted by the Brooklyn Safety Council. Catchy.

Purdy also issued a series of rules reminding drivers to treat pedestrians with respect. Sadly, this gentle effort was even less successful than his mouth exerciser—in 1932, Purdy was walking across the road when he was killed by a speeding car.

STUTTERING CURES

In the 1800s, people had curious ideas about stuttering, and many people believed the cause was a problem with the throat. A legion of inventors were ready with devices to cure stuttering.

In 1851, Robert Bates of Pennsylvania invented a two-part device. A spring clamped to the teeth, keeping the mouth open, while a clamp around the neck squeezed the throat, "thereby allowing a free passage of air through the throat and mouth from the lungs" and preventing the "spasmodic action of the throat," which was supposed to cause the stutter.

Good plan. Nobody stutters for long if they're being strangled.

Despite Bates's invention, people were still stuttering fifty years later, in 1899. That's when inventor Richard Gardner of Philadelphia stepped in with his solution. He wanted to help stutterers. At least, he said he did, although it's possible he just wanted them to shut up.

His invention consisted of two circular steel grids, held together by a loop of metal and a thumbscrew. Simply place the grids over the tongue and tighten the screw to fasten it in place. The stammerer, who now has a metal object fastened to their tongue, would then have to work extra hard to speak clearly, and the improvements would affect their speech when not wearing the device. It must have worked—nobody said a word against it.

In 1908, A European inventor named Friedrich Melzer went back to the throat. His version squeezed on the throat to "bring the vocal cords into a normal position." If that didn't work on its own, his system added electrodes, presumably to shock the troublesome vocal cords into submission.

By 1912, technology had moved ahead. It was clear that the source of stuttering was not the throat but the brain. George Peate and Thomas Beattie were two Montreal inventors who used these new insights. As they wrote in their patent, "The primary object of the invention is to provide an instrument which will perform a bloodless surgical operation in the nerve centres of the brain."

Fortunately, it doesn't involve drilling into the skull, but it is still bad. Like many of the earlier inventions, theirs is a clamp—or rather, a giant set of clamps fitting over the head. One clamp holds the tongue in place, another grips the lips, while a third holds the jaws wide open and immobile.

As the inventors saw it, this arrangement meant that the brain could only communicate through the vocal cords, and somehow this would send mental energy in the right direction to reduce the stutter. For later steps in improving the brain's nerve centres, they claimed, clamps could be removed from whichever parts of the head were currently giving the fewest problems. The strange device must have been a great conversation starter—although not too good as a conversation continuer.

None of the mechanical inventions (or surgical procedures!) designed to stop stuttering was effective. They seem more like medieval torture instruments than helpful aids.

Stutterers have included some of the most brilliant and inventive minds in history, including Isaac Newton, Charles Darwin, and Alan Turing. Some people who stuttered have found inventive ways to change their speech patterns and work around the problem. The result can be a unique and appealing speaking style. People who have achieved this include Rowan Atkinson, Marilyn Monroe, Winston Churchill, and James Earl Jones.

LITTLE BUNNY STAB-YOU

Are you afraid of hypodermic needles? Many people are, and it's a fear that starts young.

In 1967, a California dentist named Robert Smeton was concerned that hypodermic needles might be scary for children, so he invented a child-friendly version.

He designed a syringe shaped like a crouching bunny rabbit. The fluffy tail was a cotton swab that could be removed to apply local anaesthetic, then the bunny's long, needle snout was stabbed into the young patient's flesh.

We have to wonder, though—would it reduce a fear of needles or induce a fear of rabbits?

BUILD A BETTER TAPEWORM TRAP

This one is horrific. In 1854, a doctor named Alpheus Myers from Logansport, Indiana, wanted to fight the scourge of tapeworms. He invented a tapeworm trap. It was a small metal capsule with a rectangular hole in the side. A metal ring on one end of the capsule was tied to a length of fine cord.

To use the trap, the patient first had to go a few days without food. That was to make the tapeworm good and hungry. The doctor then placed a piece of cheese inside the trap, and the patient swallowed the capsule and one end of the cord—the remaining cord would still be hanging from his mouth.

Now the watching and waiting began. According to Myers, the days of starvation would cause the tapeworm to make its way from the gut up into the stomach in its quest for food. As the capsule drifted past, loaded with tasty cheese, the tapeworm would lunge for it. When its head entered the gap, a set of spring-loaded teeth snapped shut, catching the worm.

The watchful doctor would observe the movements on the string, and could reel in his catch, pulling out the string, the trap, and the entire tapeworm.

Would this work? Absolutely not. In the first place, tapeworms live in the intestines and would be dissolved in the stomach. In addition, and more seriously for the workings of this invention, a tapeworm doesn't have a mouth and doesn't wander around looking for food—the head of the tapeworm is firmly attached to the wall of the gut, and its body absorbs food directly through its skin.

Even in 1854, scientifically minded Americans mocked this invention. They said it was embarrassing that the US examiners, who had access to medical experts, could have let Myers's idea worm its way into a patent.

CHAPTER 13

MORE FLYING MACHINES

MANY of the people who wanted to imitate bird flight tried putting wings on their arms. It doesn't work. Birds have huge, powerful chest muscles for flying and humans don't.

Humans are never going to fly by attaching wings to their arms and flapping.

But there are other possibilities—like balloons. Or airships.

Or attaching wings to some pedals and pedalling.

SKY SHEPHERD

Diego Marin wanted to be the first flying shepherd.

He lived in the late 1700s, in the Spanish town of Coruña del Conde, on the river Arandilla. When he wasn't shepherding, he was inventing. He'd impressed the locals by making some improvements to a local mill. He had also created a few useful gadgets, including a tool for cutting marble and an automatic horsewhipper. When you have a track record like that, it's time to conquer the skies.

Marin spent six years building his flying machine. He used light wood and cloth for the main structure of the wings, which were shaped like those of an eagle. He covered the wings with vulture feathers he'd collected.

The wings were leg-powered—a set of stirrups was attached to an iron mechanism, which the local blacksmith, Barbero, helped him make. When he pumped his legs, the wings flapped like a fan. His hands controlled a tail flap.

On the night of May 15, 1794, he took his machine to the top of a nearby castle, aided by the blacksmith and perhaps other associates. The tower overlooked the Arandilla river. He announced that he would take a short trip and would return in a couple of days. Flapping the wings, Marin was able to get the device into the air, and he flew away from the castle and over the river.

Human-powered flight is a difficult problem even today. Modern attempts usually involve huge aircraft made of ultralight materials. Like Marin's vehicle, they are often powered by leg muscles, but the people doing the pedalling usually need the legs of an athlete. Had Marin, with his iron-framed plane, found a way past these problems more than 200 years ago?

Unfortunately, we will never know whether his progress was from pedalling or gliding, because he had travelled less than half a kilometre when one of the metal parts broke, and the plane went down.

Marin's friends hurried to the spot where he had fallen. Luckily, Marin had only received a few cuts and scrapes, but he was furious at the blacksmith for not doing a better job of welding the metal. He hurled abuse at the embarrassed craftsman. Marin told him he wanted the flying machine repaired, and quickly, so he could continue his journey.

Unfortunately, the commotion brought out the other villagers, including Marin's relatives and the village priest. They were alarmed to hear Marin's plans—it sounded to them as if he had had one lucky escape and was likely to kill himself if he tried again. It probably didn't help that his flying machine, covered in black feathers, looked like a demonic chicken. They smashed the machine, then burned it.

After that, Marin gave up on aviation. He left no plans or sketches, so people can only guess about the exact details of his flying machine, but the shepherd-inventor is celebrated today as Spain's "father of aviation"—a title that often seems to be awarded as the consolation prize for an imaginative but unsuccessful flight.

NEVER MIND IF IT FLIES—HOW DOES IT LOOK?

Most early flying machines were built for function rather than style. For example, the Wright brothers' plane was a messy assemblage of cloth, wood, and wire.

But balloons took a more elegant path. Right from the start, the French balloon pioneers applied a sense of style to their efforts. Why was there such a difference between the look of the first balloons and the first powered aircraft?

The answer, in a word, is advertising.

The Montgolfier family business was manufacturing paper. Their balloons were made of paper or combined layers of paper and cloth. They were friends with other people in the paper business, including a businessman named Jean-Baptiste Réveillon, who had made his fortune making very expensive wallpaper.

Their balloon flights, in 1783, were an exciting public spectacle, carried out in the presence of the king and other wealthy aristocrats. Réveillon said he would take care of the balloon's enormous paper coating, and he made sure it was styled and coloured with the most appealing and fashionable designs he could create. The balloon was a beautiful shade of blue, with elaborate gold decorations. The message was clear: "You loved the historic aeronautical breakthrough—now try the wallpaper!"

The Montgolfiers were cautious aeronauts. Before sending any humans up, they conducted a test flight with a sheep, a duck, and a rooster, to see if they would survive an ascent—the reasoning was that the duck, being a strong flyer, should be adapted to high altitudes; the rooster, as a poor flyer, might cope at medium altitudes; and the ground-dwelling sheep should respond to flight and height about the same as a human. The launch of the blue-and-gold balloon went well, amazing the crowd. The experimenters were surprised to discover that all the animals survived the trip. The three creatures were hailed as animal heroes and given a place in the royal zoo.

The test flight was quickly followed by a flight with human occupants—another amazing success that did plenty to keep the wallpaper manufacturer at the head of his field.

Those early balloons looked good and worked well, and it seems to have set a trend for this invention that has continued into the modern day—they remain big, brightly coloured, and often covered with advertising.

BALLOON DISASTERS

A balloon was the first aviation success, and two years later, it was followed by the first aviation disaster. But the location of this disaster is surprising: it happened not in France, which was "balloon central," but in Ireland, in the little town of Tullamore.

Details are sparse, but according to local accounts, an English adventurer visiting the town in 1785 owned a new Montgolfier-style hot air balloon. Two local men urged him to launch it.

They put fuel in the burner, set it alight, and enjoyed the balloon's ascent. It's not clear if the balloon owner was aboard the balloon, or if he was just watching an uncrewed flight from the ground.

Unfortunately, soon after the launch, the balloon drifted in the wrong direction. It struck a chimney, and the balloon's fabric caught fire. The burning balloon fell onto a thatched house, igniting the straw on the roof. Before long, the entire street was ablaze, and the fire spread fast. In total, between a hundred and 130 houses were destroyed.

Over the past two and a half centuries, the town seems to have recovered from the trauma, and their spirits have lifted. Tullamore now cheerfully hosts an annual balloon festival.

x x x

The next disaster occurred back in France, just weeks after the Tullamore crash.

The Montgolfier brothers had used hot air for their pioneering balloon, but they weren't the only team working in the field. Within days of their first flight, a competing group of balloonists launched their own creation, a balloon filled with hydrogen gas. The flight was successful, and the hydrogen balloon floated around for a couple of hours.

An ambitious science teacher named de Rozier had been on the very first human-crewed Mongolfier flight and enjoyed the accolades he had received. He wanted to make a little more aviation history of his own. He planned a truly spectacular flight—he would cross the English Channel.

But what sort of balloon would be best for such a trip? Hot air balloons and hydrogen balloons each had their merits, but de Rozier figured that if a hot air balloon was good, and a hydrogen balloon was good, then combining the two would be even better.

And that's what he built—a hybrid balloon, where one flimsy section contained a huge amount of explosive hydrogen, and another contained air heated by a raging fire.

If you think deeply, you might spot a flaw with this design that the science teacher missed.

On June 15, 1785, de Rozier was ready for the cross-channel journey. He got into his balloon, accompanied by an assistant named Pierre

Romain. The balloon rose to 450 metres, but shortly afterwards, the hydrogen caught fire, and the balloon plunged rapidly to the ground.

De Rozier didn't cross the channel, but he and his companion did succeed in marking a new first in aviation history—they were the first air crash fatalities.

THE HAWK-DRAWN CARRIAGE

It's easy to get a balloon into the air, but it's harder to make it go where you want. Steering mostly involves changing altitude until you catch a wind that will carry the balloon in roughly the right direction. Inventors had tried to find ways to steer, adding steam engines, electric motors, and propellers to their balloons, but these additions were heavy and didn't work well.

In 1887, a French inventor, Charles Wulff, thought he had finally cracked it. As he saw it, the problem with previous attempts to direct balloons was that they used parts that were "imperfectly suited to the medium in which they are placed." After all, metal machinery doesn't belong in the sky. But you know what does belong there? Birds! He proposed a "living motor."

His airship consisted of a bathtub-shaped balloon, with the balloonist riding in a basket hanging below. That much was fairly normal. But the design also had a frame running around the top of the balloon, with a circular platform fastened to the top. On this platform, four large birds were fastened by body harnesses to a circular wheel. Wulff wasn't too picky about the type of bird, vaguely suggesting "eagles, vultures, condors, &c."—you know, something big and bird-y.

A "conductor" on the balloon's top platform could steer the balloon by turning the wheel, pointing the flapping birds in any direction. Their powerful wingbeats would drive the balloon that way.

A tube, running from the top deck down to the basket, would allow the balloonist in the basket to communicate with the conductor whenever it was necessary to change direction.

Wulff explains: "It may be observed that the birds only have to fly, the direction of their flight being changed by the conductor quite independently of their own will." He clearly believed that the birds would continue flapping mindlessly, like feathered clockwork toys.

Wulff takes pains to emphasize that this was not some flying machine that is supposed to be *lifted* by birds. That would be crazy. His birds would provide only forward motion, in the chosen direction.

He doesn't explain how four birds are supposed to counteract the effects of the wind on a huge balloon. Even a gentle breeze would overpower the efforts of his feathered slaves. It's also not clear why a bird would continue to exert itself trying to fly when it just could slump in its harness and enjoy the free ride.

VACUUM AIRSHIP

A helium balloon floats in air because the helium inside the balloon is much lighter than the air around it. To put it another way, the surrounding air is heavier than the balloon, so the air pushes its way underneath the balloon, lifting the balloon out of the way and making it rise.

Helium is light, but hydrogen is lighter still, and if you put it in a balloon, it provides slightly more lift—although with some risk of an explosion.

But you know what's lighter than helium or hydrogen? Nothing! And if you could fill a balloon or airship with nothing—also known as a vacuum—that would be the ultimate in lifting power. The idea of a "vacuum balloon" has been around for centuries.

In 1899, inventor Arthur de Bausset tried to get money from American investors to build an airship based on this principle

The airship he designed was a monster—around 200 metres long and 42 metres wide. It was to be shaped like an enormous cylinder, with cones at each end to make it more streamlined. Instead of being covered with fabric like the later Hindenburg, the outside of de Bausset's ship was made from plates of steel, each just over half a millimetre thick, welded together to make an airtight container. The inventor had confirmed that this thickness of steel could easily contain the vacuum without breaking.

His ship had enough lifting power to carry 180 passengers (or the equivalent weight in cargo) and would travel up to approximately 95 kilometres per hour.

Just as wood floated in water, the inventor claimed his airship would float in air. Forward propulsion would come from electric motors.

De Bausset asked skeptical investors to consider if his proposal was really so far-fetched. He asked: "Is it not worthwhile to make trial of a scheme which, if successful, would revolutionize transportation?" Then, he continued, in the style of a supervillain: "Would not the holder of a patent giving him the monopoly of such a means of transportation command the wealth of the world?"

De Bausset's ideas were published in newspapers of the day, and it was reported that American investors were interested in the plan. If that's true, they didn't pony up enough money to build the craft. That's just as well, because his ship absolutely would not have worked. It's true that a sheet of thin steel can easily separate a vacuum from the air around without puncturing or breaking. But so can a large soda bottle—and if you suck the air out of a soda bottle, it will crumple and collapse under the crushing pressure of the atmosphere. That's what would have happened to de Bausset's airship: even if it had remained airtight, it would have been squashed.

In recent years, scientists have returned to the problem of the vacuum balloon. It might be possible to build one using advanced materials, like boron carbide ceramic or films of diamond-like carbon. Perhaps one day it will happen. After all, the only part of a vacuum airship that's hard to build is the outside. As for the inside—there's nothing to it.

FLYING SHELVING UNIT

If you look up pictures of weird flying machines, you may come across an image of the "New Flying Machine" by W. O. Ayres, which looks as if it would have absolutely no chance of leaving the ground.

But the person who designed the machine was no fool. William Orville Ayres was a professor at Yale, an expert on diseases of the nervous system. In addition to being a medical doctor, he was an expert in fish. He was also an authority on birds, so it seems like he should have known a thing or two about flight. Why would a man with his education and knowledge propose such a ridiculous machine?

The answer is that he probably didn't.

Ayres liked to share his thoughts with the world, and since there weren't as many scientific journals in the 1800s, he sometimes published articles in newspapers and magazines aimed at the general public.

In 1885, *Scientific American* reported on many of the latest inventions and technologies. Ayres wrote a short piece for the magazine, describing how a flying machine might work.

Ayres explained his flying machine in the text. He envisioned something like an aerial bicycle. The pilot would sit inside a tubular metal frame, with a number of fans at the top and sides, using his legs to drive a pair of pedals connected to the whirling blades. On its own, that might not be enough to get airborne, but the vehicle would also be equipped with two cylinders of compressed air. The cylinders would power a small engine, working like a steam engine, but without the weight of coal and water.

He seems to be describing something like a pedal-powered helicopter with several large rotors. It probably wouldn't have worked, but the idea wasn't crazy either. However, the magazine needed a picture to go with the story and seems to have passed the description to one of its regular illustrators.

It was the artist's job to interpret an inventor's description and make it look real, but their choices were sometimes odd. For example, in the same issue as Ayres's article, an artist illustrated a new toy. It was described as a locomotive with rockers like a rocking horse. In the accompanying picture, a small boy is shown riding the vehicle the way a child would have ridden a rocking horse, thrashing his locomotive with a horse whip.

The illustration for Ayres's idea was even more bizarre, with odd proportions. It shows a rosy-cheeked young man with windswept hair, sitting inside the frame of a cheap wire shelving unit, supported by a set of ridiculously tiny propellers. He absent-mindedly turns an infant-sized set of pedals, apparently connected to the rotors by a loop of yarn. The device might work to create a breeze in a room, but there's no way it would fly.

Ayres saw the illustration before the article was published. He seems to have been uncomfortable with the way his ideas were presented but perhaps did not want to appear rude or ungrateful for the artist's efforts. He tried to distance himself from the drawing, writing tactfully: "The plan and form which we suggest is not designed or expected to be by any means exclusive . . ." and he goes on to explain that "the propellers may be made to present a much greater extent of surface than the artist has drawn . . ."

His disclaimers didn't matter. The picture was ridiculous and completely undermined the reasonable points he was trying to make. Today, the crackpot flying machine, with Ayres's name attached, is all over the internet, as well as on bags and T-shirts, as an example of the crazy ideas people used to have about flying.

If you want to be remembered as an illustrious inventor, make sure you get a good illustrator.

EIFFEL TOWER SHELL DROP

The "egg drop challenge" is often given to aspiring engineers. It involves designing a container for an egg that would allow it to survive a drop onto a hard surface from a great height—for example, the top of a tall building.

A French inventor named Charles Carron came up with a machine that solved a similar problem—except that instead of dropping eggs, he proposed dropping a group of fifteen thrill-seeking passengers 300 metres from the Eiffel Tower.

The device resembled a huge, pointed artillery shell, pointing downwards. According to his plans, the shell weighed 10 tons, and its interior consisted of a lounge-cabin area resting on a series of collapsible cones. Access to the cabin was through a door in the side. The floor was built like a huge mattress, resting on 50-centimetre springs. An electric light fixture hung overhead. An illustration showed a circle of fifteen well-dressed men and women on padded armchairs in the lounge. In a strange oversight, the device lacks windows—the passengers gaze outwards at a blank wall.

Carron was a blue-skies thinker. To make his plan work, he intended to build a large body of water directly below the Eiffel Tower—a wide pond which narrowed into a deep hole, like a well.

He intended to hoist the shell up to its maximum height, then drop the device, along with his paying customers. If all went according to plan, the object would fall for several seconds, accelerating to around 270 kilometres per hour. It would then strike the water, which would slow its descent, while the springs and layered construction reduced the impact for the occupants. When it hit the bottom of the pond, air inside the walls of the shell would cause it to bob to the surface again. The sensation-seekers could then leave (or their remains could be otherwise removed), and a new batch could climb aboard, at twenty francs a head.

Fortunately, the device was never built.

CHAPTER 14

YOU'RE NOT GOING OUT IN THAT?

THE words we use about clothes are telling.

If someone shows you a jacket they've "designed," you have an idea of what to expect. You might admire the jacket's cut, its colour, the way it flows over the body.

But if they show you a jacket they've "invented," the expectations are different. Now it's about how the jacket works—is it bulletproof? Does it float? You expect this article of clothing to do something new.

There's no reason a great clothing idea can't also look good, but in practice, invented garments are often designs with no design sense.

LOVELY GLOVE

A British invention from 1990 is focused on a problem with gloves—they pose a barrier to holding hands. Some couples may choose to hold hands while keeping their gloves on. They experience only glove-on-glove contact. Others may choose to remove their gloves and feel each other's hands—but they will also feel the cold. As inventor Terence King rightly says, "This is unsatisfactory and indeed unromantic."

That's how King came up with his invention, which is oddly specific. It is a shared glove for couples who want to hold hands in the cold—a couple who are desperate for their palms to touch but want to keep their fingers aloof. The glove has eight interconnected fingers and two thumbs, providing splendid and exclusive digit coverage for each individual wearer. Each person's palm, however, is in flagrant, naked contact with that of their partner.

A mitten version works in a similar way, maintaining palmar proximity, but keeping the fingers and thumbs of each hand in their own separate compartments.

The inventor explains that the glove "may also be so sized as to accommodate and fit the respective hands of a mother and child." That is, a mother and child who want to touch palms, but can't stand the idea of intermingled fingers.

The patent application describes other options, too. For example, the two halves of the glove can be different colours, so that, to onlookers, the wearers appear like two ordinary people holding their gloved hands together, rather than two people with strange ideas about skin contact.

SHARED OVERCOAT

The idea of a shared garment can be taken further. Say you're walking out with a friend, and it starts to rain. You wisely wore your raincoat, but your optimistic and impulsive friend went outside without one. Now you will feel awkward, protected by your coat while your friend gets soaked. Two people can share an umbrella, but they can't share a coat.

Or can they?

That was the realization that struck inventor Howard C. Ross in 1953, and it led him to design his revolutionary double coat.

In its normal form, it looks like an ordinary overcoat—if, by "ordinary," you mean a baggy garment with a strange collar, three rows of buttons, and an incongruous zipper that runs all the way down the back.

But the coat can be expanded to twice its width by undoing the zipper and rows of snap fasteners, then extending a set of fabric panels. Two people will then fit inside it. Both their heads emerge side by side from an extra-wide collar, and they each get one coat sleeve to themselves, while their other arm occupies the shared space inside the coat. (They can expect comments like "What's going on under there?" from amused passersby.)

If the wearer is out with a lady friend, and it suddenly starts to pour, he can set to work adjusting the various panels and fasteners. In a few short minutes, the coat will be large enough for the two of them.

His friend then faces a difficult decision. Will she be more embarrassed by standing in her drenched jacket and skirt, or by climbing inside a man's coat and becoming part of his two-headed monster costume?

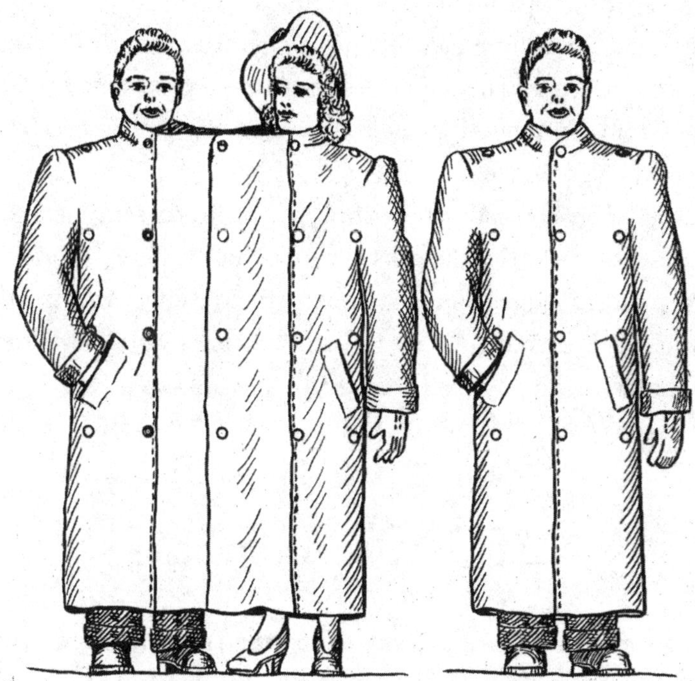

The illustration on the patent shows a woman who made the second choice and is perhaps regretting the decision. Even as a line drawing, it is clear that her girlish dreams of romance have vanished forever. Her expression seems to say, "What am I doing wrong with my life? How did it come to this?"

Ross's overcoat displays a fundamental paradox. It is an overcoat that can be shared with friends, but anyone who wears it by choice may not have any friends to share it with.

NOSE TO THE GROUND

In winter, you can warm your cold hands by blowing on them. Unfortunately, that method doesn't work to keep your feet warm—or it didn't, until American inventor William Tell Steiger considered the matter in 1877 and came up a pair of William Tell overshoes.

Seiger had carefully measured the temperature of his own breath using a thermometer, and he didn't like the way the precious heat from the lungs was going to waste on a cold day. That warm air could be better used to heat his frozen feet.

His invention consists of a rubber mouthpiece strapped over the mouth and attached to two lengths of rubber hose, which run down to two shoe sole inserts. Steiger claims to have tested the effectiveness of his device. If the user breathes out, they will send a blast of warm air to the feet.

On the downside, if they breathe in, they will send a blast of smelly foot air to the nose.

A SEAT FOR YOUR SEAT

Everyone appreciates clothes that are designed for comfort, and comfort isn't just about staying warm. Sometimes you just want to sit down.

This 1993 patent from British inventor Michael Bayley combines clothing with furniture. It's a seat that can be worn on the body. When you sit, you can unstow the seat and turn any surface of suitable height into your very own padded chair. If no other surface is available, you can just sit on the ground, knowing that you are fully protected from dirt and damp.

The problem with the design—as with many clothing inventions—is that it is physically cumbersome and displays a total lack of fashion sense. In the stowed position, the seat is clipped to a belt. The wearer looks as if they have a chair cushion hanging over their buttocks—which, basically, they do.

POCKET FOR WOMEN

A common complaint about women's clothes is the lack of pockets. Clothes for men have good-sized pockets and plenty of them. Pants alone often have four pockets, providing a place for wallet, keys, loose change, and a Swiss Army knife with forty tools. Women's pants frequently have no pockets or pockets that are ridiculously tiny.

It's an old problem, and even long ago, some women were doing what they could to solve it. Harriett Lucinda Peterson, a Canadian living in Portland, Oregon, played her part in the struggle back in 1919 by designing a "secret pocket and hose supporter." It was a small pocket with a clip at the top. It attached to an elastic garter—the old-fashioned kind that held up a woman's stockings.

Using Peterson's invention, a woman could easily add a pocket beneath her skirt—or even multiple pockets—providing useful storage room for all the items that would normally have to be carried in a purse.

It still didn't quite compete with the men's pockets, though. A man could smoothly reach into a pocket to pull out a coin or watch, but for a woman to remove anything from her hose-supporter pocket, she would need to hoist her skirt and go fumbling through her underwear.

HAT TRICK

In 1896, most men wore hats. It was good manners to tip your hat as a show of respect. But it was a nuisance to be constantly reaching for your hat—especially if your hands were doing something else at the time, like carrying groceries or adjusting your silk tie. Inventor James Boyle of Spokane, Washington, had a solution—a clockwork-powered self-tipping hat.

The mechanism was hidden under the hat and gripped the wearer's head with several curved arms. The hat was attached to the top of the device. Most of the time, it appeared like an ordinary hat on an ordinary head, but if the wearer nodded his head, the movement would cause a pendulum to swing forward, setting the mechanism in motion. The hat would then rise from the head, tip forward, then swing around flamboyantly in a circle and set itself back onto the owner's head.

In the inventor's words, the purpose of the hat was "for automatically effecting polite salutations." Some might see the greeting as sending an additional message: "I can't be bothered to tip my hat to you, so I'll have a machine do it for me." At a minimum, it tipped people off to the wearer's eccentricity.

Boyle's true colours come out when he describes the advertising opportunities offered by his mechanical hat. You could monetize your good manners by putting an advertisement on a small sign or placard attached to the hat, and greet people on the street, "the novelty of its apparent self-movement calling attention to the hat and its placard."

KEEP A COOL HEAD

Having to wear a hat all the time could make a man's head awfully hot. One obvious solution was to remove the hat. Inventors preferred another idea—the hat ventilator. The sheer number of these inventions is incredible—it must have been invented more often than it was sold.

Most designs required an oversized hat. Some installed a band inside the body of the hat, to create a space between the head and the hat. An 1875 hat ventilator design pressed out the sides of the hat, creating a gap for air to move. An 1880 design lifted the hat slightly off the head.

Year after year, the ventilators kept coming. It seemed that all men wanted to do was remove their hats at the first opportunity, and many inventions were designed to lift the hated hat away from the scalp.

A different approach was to add holes to the hat, so air could blow through it. There were many of these inventions, too, often designed to make the shameful ventilation holes less visible.

A British design from 1898 added a hidden pocket to the hat, to create a secret gap where cool air could flow.

An inventor named Walter Montgomery proposed rubber inserts, coloured to match the hat, which could be inserted into the fabric to create an air hole. John Wilson went further—in addition to a ventilation hole on top of the hat, a sliding mechanism discreetly opened up three more holes in the front.

In 1892, William O'Connor went further still, designing a hat that would have suited the giant rabbit Harvey. It looked like a normal hat, but when the heat buildup became too much, you could press a mechanical button and open two flaps in the top, perfect for rabbit ears.

O'Connor returned to the theme in 1889, with a better version—again a normal-looking hat most of the time, but when nobody was around, the wearer could press the switch and lift up the entire top section of the hat for ventilation. Again, why not simply remove the hat?

In 1906, William Webb patented a hat with a rubber grille around its side instead of a band—a series of slits allowed the movement of air. In 1908, William Miller did something similar using aluminum strips.

The most extreme hat ventilators used mechanisms to move air inside the hat.

In 1860, James Jenkinson designed a ventilator that would only fit inside a tall top hat—although that wasn't a problem, according to Jenkinson, because such a hat was "in common use by all classes of society and in every civilized country." His invention was a set of four round flaps on springy arms placed inside the hat. As the wearer walked around, these flaps would move up and down, and create "a gentle current of air . . . in the interior of the hat whereby the head is cooled, and the foul air is expelled."

An 1890 patent from Albert Lee Eliel of Illinois involved putting a clockwork fan inside the hat, venting the hot air out through the top. He doesn't say how long the fan was expected to run, but you probably wouldn't get much cooling time from each windup.

In 1914, Lydia Tyler took an interesting approach. In addition to holes on the side of the hat, the hat contained a swinging rectangular flap hung on the inside from metal hooks. As the wearer moved around, the flap moved back and forth, fanning the wearer's hot head—and perhaps squeaking.

Although the fashion for men's hats faded after the 1960s, the ventilated hat patents have kept on coming. Modern versions are usually for baseball caps, often with attached electric fans powered by solar cells or batteries. Today, as in centuries past, they serve a triple function—they keep the sun off, keep your head cool, and provide something for people to laugh at.

WOMEN'S HAT SUPPORT

Of course, women also faced hat problems. In 1912, ornate hats were fashionable, but big hats were heavy and hot. If they were large, they might catch the wind and blow off.

A Louisiana resident, Arthur Munchausen, was ready with a solution. His hat support comprised two metal struts, one on each shoulder, which supported the hat from the sides like hat crutches and were fastened to the woman's shoulders by pretty bows.

The inventor said the hat could be kept slightly above the head, which would allow a free circulation of cooling air. Using his supports, a woman could wear a huge sun-blocking hat and would not have to worry about carrying a parasol, or she could keep her hands free in a rain shower, wearing a rubberized hat instead of holding an umbrella.

His system worked for more decorative ladies' hats too. Supported by Munchausen's wire frame, the hat would remain upright and elegant, proudly displayed as the powerful fashion statement it was intended to be—although she would have to remember to keep her head pointing straight forward, otherwise, it would be facing in a different direction than her shoulder-supported headwear.

DATE HAT

How frustrating it must have been in the 1950s for a boy to ask a girl out for a date only to discover that the evening he's picked is one when she's busy.

In 1958, Canadian inventor Edward T. Oliveira focused on this complex problem of dating the opposite sex, and he solved it—at least, in his own imagination.

His patent was for a "date hat." It was worn on the back of a girl's head and was marked with a clock dial design. Its segments showed times and upcoming days of the week. Adjustable hands on the top of the hat could be pointed to times and dates that the girl particularly wanted to fill.

A little flap is hidden under the various days. When a day has been booked, the flap is unfolded past the edge of the hat, displaying a sign that says "Taken." Perhaps not the best choice of words.

The inventor's utopian dream seems clear. Once the hats were installed on every girl's head, the messy world of dating could be streamlined for everyone. At last, boys who were confused by the complexities of relationships and romance could find the clarity they yearned for. He could follow a girl around, and a quick glance at the back of her head would show precisely and unambiguously when she was available to be dated.

The girls, for their part, could ensure that any free evening would be quickly filled with a date with a random boy of their choosing. Girls could also have fun comparing their hats, seeing who had a week filled with dates and who was a sad, lonely reject.

That was the grand plan, and yet today this headwear is only a forgotten vision of what might have been. For some reason, the date hat never received any love.

WOODEN HATS

Sometimes an invention goes nowhere but leads to other opportunities for the inventor.

Around the year 1777, a twenty-three-year-old Scottish inventor named William Murdoch wanted a job in engineering. He thought he'd try his luck getting a job with James Watt, who was making big money in the steam engine business. Murdoch grew up in a little village named Lugar, and Watt's birthplace wasn't far from Murdoch's, so he expected to get a sympathetic hearing, at least.

Even a naive country boy knew you couldn't go calling on important businessmen without a top hat. Murdoch didn't have one and couldn't afford to buy one. But why buy an expensive hat when you can just make your own?

In those days, many men's hats were made from silk or felt made from beaver fur. Murdoch wanted something more durable. He decided to make his top hat from a block of wood, turned on a lathe. Unfortunately, he didn't have a lathe either. But why buy an expensive lathe when you can just make your own?

He made a lathe and modified it, so instead of cutting a round hole, it would make an elliptical one, better suited to a human head. Then he found a piece of wood and made a plausible hat from it. He painted it to resemble the sort of hat you'd buy from a hatter.

Finally, dressed for business, the inventor walked to Watt's office. This was more of an accomplishment than it sounds. Watt didn't live in Scotland anymore. His office was in Birmingham, nearly 450 kilometres away.

After a very long walk, Murdoch finally arrived at the firm of Boulton & Watt, ready for his fireside chat with the great engineer. Unfortunately, Mr. Watt was out, and instead his meeting was with the co-owner, a sophisticated Englishman named Matthew Boulton. Predictably, it didn't go well—as the nervous lad sat there, fidgeting with his hat, Boulton said basically, "Thanks awfully for dropping by, old chap, but we don't have any vacancies at present."

But before Murdoch could leave the office and begin the long walk home, Boulton had to ask about the lad's strange-looking hat. It wasn't silk, and it wasn't fur. What on earth was it?

Murdoch explained that it was made from wood, and after some prompting from Boulton, he explained how he'd made the lathe to produce it. Boulton was an engineer himself, and he was amazed at Murdoch's ingenuity in building a lathe that could cut an elliptical hole, all to save money on a hat.

Murdoch was hired and eventually became the company's chief engineer. He devised many inventions that the company patented. But the wooden hat concept wasn't one of them—he was allowed to keep that for himself.

CHAPTER 15

THE RIGHT CHEMISTRY

AFTER inventing a wooden hat, and getting a job working for James Watt, William Murdoch came up with many ingenious ways to improve the company's steam engines.

Murdoch was an unusual character. Although his main work involved designing and fixing machines, he became increasingly interested in what we would now call industrial chemistry. His first chemical invention made him thousands. His second made other people millions.

FISH BEER

Beer and fish don't seem like a natural match in a drink, or a promising area for an inventor, but this was the odd specialty that William Murdoch decided to explore, when he had time off from fixing James Watt's steam engines.

At some point in the distant past, beermakers had discovered that sturgeon bladders were a great way to store beer. That's partly because the sturgeon is a huge fish, and you can fit a lot of beer in its swim bladder. But storing beer inside this fish's innards had an unexpected benefit—you could put cloudy beer into the pouch, and when you poured it later, it came out clear. That's because the inside of the bladder is mostly gelatin, and when the clouds of floating yeast are exposed to it, they form clumps and settle to the bottom, leaving a beautiful clear beer behind.

By the 1700s, most brewers were storing their beer in kegs and bottles rather than fish skins, but they remembered the fish-skin trick and routinely added pieces of sturgeon bladder to their beer. It worked just as well—the yeast would drop to the bottom in a gooey clump, and the brew would quickly go from cloudy to clear. This sturgeon-bladder jelly is known as isinglass, and it used to cost a fortune. That's because the isinglass had to be shipped from Russia, where the sturgeons (which are endangered today) were killed for their caviar. In the 1700s, isinglass sold for twenty-five shillings a pound—it wasn't quite worth its weight in gold, but it was worth more than its weight in silver, and brewers went through a lot of it.

Murdoch didn't see what was so special about the sturgeon. Isinglass was just an animal glue, and he figured it should be possible to make a cheaper version from regular British cod. He carried out a few experiments, using his rented accommodation as a workshop. According to one account, his landlady went in to check on the luxurious apartment, and when she saw that her tenant had pinned dead fish to the walls, she evicted him.

Still, it was worth the inconvenience. Murdoch successfully figured out a way to make high-quality isinglass from cod. A group of London brewers paid the inventor the princely sum of £2,000 for the right to use his invention. Isinglass is still used by modern brewers. For years, the glue was also used as the sticky gum on stamps and envelopes, and as an ingredient in jellies and other wobbly desserts.

BAKED COAL

William Murdoch worked for years for James Watt's company. He came up with a series of inventions, most of which were patented under Watt's name—because Watt paid Murdoch's salary. A century later, Thomas Edison had a similar arrangement with the creative people he employed.

It wasn't all inventing for Murdoch, though. Most of his time was spent setting up and repairing Watt's steam engines for coal mines and factories. He was spending a lot of time watching coal burn. But an inquisitive mind is never short of questions. Murdoch became curious about exactly how coal works.

People had known for a long time that if you bake coal without burning it, it changes—a gas streams off the top of it, a dirty tar oozes from the bottom, and what you're left with is a dark grey material called coke. It is almost pure carbon, and if you light it, it acts like super-coal, burning at a very high temperature.

But Murdoch was more interested in the waste products—especially that escaping coal gas. The gas was flammable, and when he tried burning it, he was surprised at the results.

Say "gaslight" today, and people think of a red flag in relationships, from an old movie where a villain carried out psychological manipulation by adjusting flickering yellow gas lights. In real life, gas light was steady and bright. Murdoch was impressed by the white flame it produced. It also didn't stink like an oil lamp, and it was much cheaper to run.

He realized that this coal gas was one waste product that really shouldn't be wasted. If you could build containers for storing the gas, then run pipes to homes, streets, and factories, it could light up a whole city.

Thanks to Murdoch's work, companies everywhere were soon using the new gas, which became known as "illuminating gas." In Paris, gas lights lit up the streets at night. People had been used to going home after dark, but with the new lighting in the city, they could socialize or go to restaurants. It was the first real nightlife. Other cities soon followed. After lighting the streets, the new lights were installed in homes, offices, and factories. In theatres, it became easy to light the entire stage, and the brightness could be easily adjusted by turning a gas valve.

Of course, nothing is perfect, and there were two problems with the new gas. One was that, as with any flammable gas, a leak could cause a fire or, if enough gas accumulated, a huge explosion. You might not notice a leak because the gas didn't have a strong smell.

Another reason you might not notice a leak is that the escaping gas was quite likely to kill you. The illuminating gas was mostly a mix of two other gases. One was hydrogen. That's the stuff the Hindenburg was filled with—extremely flammable but otherwise not that dangerous. The other gas was carbon monoxide.

Today, we associate carbon monoxide with exhaust fumes, so many people are surprised to learn that it is flammable. It burns well, with a very white flame. But, of course, it is also extremely toxic. Red blood cells mistake the gas for oxygen, latch onto it, then fail to release it again. It is a silent killer. If an unlit lamp leaked gas into a bedroom, the gas could easily kill a person as they slept.

Still, despite the risks, most people viewed gas lighting as an improvement in their lives. The new lighting made an even bigger splash than Watt's steam engine and might have made Murdoch a wealthy man or brought in an even bigger fortune for James Watt's company. It didn't happen.

Although Murdoch urged his boss to patent the new gas lights, Watt dragged his feet and only made half-hearted efforts to introduce the technology. He may have been too busy fighting to protect the patents on his steam engines to want to spend time on a whole new business. Or perhaps Watt just wanted to maintain a tight control over his lead inventor—a little gaslighting over gas lighting.

In the end, Murdoch and Watt did nothing to protect the gas lighting invention. Other companies installed gas into homes and businesses and made enormous profits from it.

If James Watt failed to patent a good invention, he also failed to patent a very bad one—in fact, it must rank as Watt's worst idea. He suggested that the coal gas could be useful as a therapy, and that it might act as an "antidote" to an excess of oxygen in the blood. It was certainly an antidote to staying alive. Inhaling a few deep breaths of carbon monoxide is a procedure with a high chance of killing the patient.

Ironically, Watt may have been using this "therapy" accidentally for much of his life—he suffered from terrible headaches, which can be a side effect of inhaling the carbon monoxide released from burning coal.

A GHOSTLY GLOW

Carl Auer von Welsbach was an Austrian chemist and inventor. He was fascinated by a group of silvery elements called "rare earths." Despite the name, they're not that rare, but they're also not that obvious, and it took people a long time to discover that they were there at all. Welsbach himself discovered two of them.

Unusually for a scientist, Welsbach was good at turning his discoveries into practical products for everyday use. For example, after working with the element cerium, he invented an alloy that made sparks when you hit it against steel. He started a company to make the metal, and it is still used as the "flints" in lighters today.

Welsbach was also interested in gas light. In the 1890s, lighting with gas was wasteful. It took a lot of gas to light a large room, and although the gas burners produced impressive levels of light, they made even more heat. One scientist said that lighting your house this way made as much sense as burning down your house to cook your dinner.

Welsbach wondered if there was a way to use the wasted heat to create a more efficient light. Some theatres used a lamp that made a brilliant light by shooting a flame at a cylindrical block of quicklime. The cylinder became white hot and gave off a brilliant light. A limelight was quite large, so it was used in theatres as a spotlight—which is why, even today, a person getting all the attention is still said to be "in the limelight."

Making a chunk of rock glow white required an extremely hot flame—usually a jet of hydrogen and oxygen. Welsbach wanted to get the same sort of result with a regular flame. That was possible if he put the available heat into something lighter in weight.

He came up with a strange addition to the gas lamp. It was called a gas mantle, and it looked like a small, knitted skirt woven from cotton. It was soaked in chemicals made from his favourite rare earth metals.

When the gas mantle was placed over a flame, the cotton immediately burned away. What was left behind was a network of ash. The ash mantle looked identical to the cotton skirt, but it was like a ghost of the original—extremely lightweight, and so delicate that it would crumble at the slightest touch. There was so little of the mantle that a regular flame turned the ash white hot.

With Welsbach's mantle installed, any gas flame became ten times brighter. That was a huge improvement. Suddenly, with this small modification to the lights, a room or street could be brightly lit with a tenth of the gas.

Welsbach had come up with an amazing invention, but it arrived too late. By the time his new mantles came out in 1890, Edison was busy lighting up streets and homes with electric lights. The new electric bulbs weren't as bright or as white as Welsbach's mantle lamp, but they were easier to use. The mantle lamps slowly disappeared, and today they're rarely seen, except in some propane-powered camping lanterns.

The quick decline of the new gas technology was unfortunate for Welsbach, but he could see the writing on the wall for gas light, and he was nothing if not versatile. He experimented with electric light filaments and found a way to make them out of the precious metal osmium. His bulbs lasted much longer than the ones made by Edison, and Welsbach's light bulb company eventually became the company Osram, which became—and remains—one the world's largest manufacturers of light bulbs.

CHAPTER 16

CHEMICALS TO DYE FOR

SOME inventions are accidental—an inventor or scientist muddles along trying to solve one problem and ends up solving a different one.

People had made big money selling the gas that comes off heated coal, and their success inspired others to take a closer look at another waste product from coal—the tar that drips from the bottom.

Coal tar didn't look like a promising area for a science-minded inventor, but the chemicals in that sticky brown goo made a fortune and turned organic chemistry into a multi-billion-dollar industry.

Most of the biggest success stories were accidents—this was one group of inventors who didn't have a clue what they were doing, or even what they'd done.

THE PERKS OF PERKIN'S PURPLE

Have you ever started on one project only to be completely distracted by something else? While most people will say success requires discipline and focus, letting your mind go where it wants can sometimes bring fame and fortune. That was certainly true for a young chemistry student named William Henry Perkin.

In the 1800s, scientists were making breakthrough discoveries in chemistry, trying to find ways to use their skills to solve social problems.

A big problem for the British Empire was the tropical diseases in all those hot colonies. Malaria was one of the worst—it made millions of people sick. A drug called quinine could prevent the disease, but it had to be extracted from the bark of a Peruvian tree, so it was expensive. Some scientists believed it might be possible to create the life-saving drug in the laboratory. Whoever achieved that would probably make a lot of money.

In 1856, William Perkin was an eighteen-year-old chemistry student. He had heard about this quinine problem and wanted to see if he could solve it. He carried out experiments using coal tar, the sticky brown goo that is extracted from coal. Coal tar and quinine were both organic chemicals, meaning they were made from carbon, and in theory, the chemicals in tar could be transformed into quinine. Perkin gave it a try, but he didn't make any headway on a malaria cure. However, as he was cleaning out the dirty test tubes with alcohol, he was surprised to find his cleaning cloths turning purple.

Perkin was impressed by the brilliant colour. He had an interest in art, and he thought the new chemical might make a useful dye. The purple looked like the colour of a little flower called a mallow. In French, the flower name is mauve, and that's the name he gave to the new colour. He named the chemical mauveine. He had discovered one of the very first synthetic dyes.

We're used to being surrounded by colours today, but in the 1800s, being able to make bright colours from cheap chemicals was a big deal.

Perkin shopped his discovery around to dyeing businesses—this was the middle of the Industrial Revolution, and factories were producing huge amounts of fabric, which meant they needed large amounts of dye.

The industry experts took one look at his samples, loved-loved-LOVED the colour, and wanted to use it. Perkin smelled profits and took a gamble—he quit school and borrowed money from his father, then used it to set up his own dye factory.

After he started selling the new dye, Perkin had a lucky break—Queen Victoria started wearing a purple colour very similar to Perkin's mauveine. Before he knew it, mauve was the colour of the year. People hardly ever used the word before, but now "mauve" was on everyone's lips—and on everyone's hips. It was even used on postage stamps—a six-penny stamp was printed using Perkin's new dye.

The dye was a beautiful colour, but it tended to fade when exposed to light—and it would keep fading to almost nothing. That may be why, when people talk about mauve today, they imagine a pale colour, a pinkish lilac. When it first came out, mauve was vivid and attention-getting. The dye was a true "fading violet."

Despite the issues with fading, Perkin did very well from the sales of his patented dye. His dye business was a success, and he also invented ways to make other colours from coal tar. His colour experiments made him a wealthy man.

But what about the quest for quinine? Perkin had failed there, but he didn't need to feel too bad about it. Creating quinine from scratch seemed like it might be an easy chemical problem in the nineteenth century, but scientists kept working on it for 150 years before they finally figured out a theoretical process in 2018. Even today, quinine is still produced from Peruvian tree bark.

A TINT OF TNT

After seeing how much money William Perkin made from his new mauve dye, other chemists wanted to get in on the act. Germany had a great many good chemists, and they soon started competing for their share of the dye pie.

In 1863, a German chemist named Joseph Wilbrand tried to make a new yellow dye. He started with the same family of coal tar chemicals Perkin had been working on.

One chemical in this group produced a pleasant yellow colour, but it wasn't impressive enough to be the new mauve, so Wilbrand moved on. It's too bad, because although the chemical he'd found was only so-so as a dye, it could have made him a fortune for one of its other uses.

Wilbrand's discovery was trinitrotoluene, better known as TNT.

Then, as now, explosives were used for warfare, mining, and demolition. One of the most widely used explosives was nitroglycerine—a yellowish, oily liquid. It could produce a powerful explosion, but it was incredibly dangerous to work with. It could explode if it was hit by a spark, or if you dropped it, or shook it, or bumped it, or even exposed it to sunlight. Workers were often killed by accidental explosions.

Four years after Wilbrand had given up on his yellow dye, the Swedish chemist Alfred Nobel had the idea of mixing nitroglycerine with a kind of clay and wrapping it in a cardboard tube. This made the nitroglycerine less sensitive. He called it dynamite. Now the explosive could be moved around more safely and was less likely to go off by accident. The invention made Nobel embarrassingly rich.

If Wilbrand had only realized what his yellow dye was capable of, he would have made an even bigger fortune than Nobel, because anything dynamite can do, TNT can do better. It is not only more powerful than dynamite, but also much safer. It can be handled, dropped, or even cooked without causing a blast. It was exactly the thing the world was looking for.

But it was decades before other scientists took another look at TNT and discovered that the obscure yellow dye was an explosive—talk about a colour that really pops. After that, the chemical was produced in Alfred Nobel's factories, and Nobel made money from TNT as well as dynamite.

TNT was never used commercially as a yellow dye, but it worked that way accidentally. During World War I, munitions workers, who were mostly women, had to handle the substance as they assembled bombs and shells. It turned their skin yellow, giving them the nickname "canary girls."

PLAYING CARD BOMB

As we've seen, the chemicals from coal tar can be used to make dyes. As we've also seen, they can be used to make explosives. A man named William Kogut found a grisly but inventive way to combine the two.

Kogut was on death row in San Quentin Prison. It was October 1930, and the Polish immigrant had been sentenced to death for killing a woman named Mayme Guthrie, the owner of a rooming house.

Kogut had been drunk when he committed the murder. When he was arrested, he said he had no memory of what he'd done, but he believed the accusations and said that if he'd killed her, he deserved to die. Perhaps he was still drunk when he made the statement, because later, he had no memory of saying that either.

He was sentenced to be hanged. Imprisoned in San Quentin, Kogut awaited the results of a final Supreme Court appeal, but he knew it was unlikely to save him. Kogut was determined to end his own life rather than be hanged. Death row inmates don't have access to deadly weapons, so Kogut decided to invent another method.

On October 20, prison inmates heard a huge explosion from Kogut's cell. When guards rushed to Kogut's cell, they found him close to death, with terrible head injuries. He was taken to the prison hospital, where he died a few hours later.

It appeared he had taken a metal tube from his bed and used it to create a pipe bomb—or perhaps it was intended as an improvised shotgun—which he held to his head. Investigators found that pieces of wet, torn-up playing cards had been stuffed inside the tube, and it looked as if it had been heated on Kogut's oil stove until it went off. Kogut left a suicide note, saying that he had arranged his own death, and nobody else was to blame.

Clearly, Kogut had invented some kind of bomb, but how?

Some said it was just a matter of steam pressure. If you seal a tube at both ends, then boil water inside it, it will eventually explode, and the plug flying out can do a lot of damage.

But others believe Kogut's method was more sophisticated. Kogut knew that the red dyes used in playing cards were closely related to the chemicals used in explosives. Prisoners were allowed playing cards in their cells, so Kogut accumulated twenty packs and carefully tore out the red markings from the cards. He mushed the pieces in water to make a chemical paste. He added a few more chemical ingredients—a couple of everyday substances that could be found and swiped from prison medical supplies—and mixed them into a mixture that had the explosive power of nitroglycerine. Finally, like Captain Kirk fighting his lizard opponent, he stuffed the ingredients into a tube and waited for an explosion.

A similar recipe, known as a "hearts and diamonds bomb," appears as one of the improvised bombs in the controversial 1971 counterculture *Anarchist Cookbook,* although how Kogut obtained this knowledge in 1930 is a mystery. He was a logger by trade, but using this method suggests a knowledge of some advanced chemistry. Did he have a secret past?

It's also possible that Kogut tried to make an impractical bomb from playing cards using misguided information he'd heard at the gambling table. In this version of events, the bomb would not have worked from a chemical reaction, but he "got lucky" and instead of being killed by a chemical reaction, he was killed by an unexpected steam explosion.

We may never know the truth of the matter. It's hard to duplicate his methods because cards today use different inks and coatings. Of course, Kogut didn't share the details of his invention—it was only going to be used once—but from the suicide note, he seemed confident that it would work. People are still debating how he did it.

Perhaps a clear answer just isn't in the cards.

BAYER HEROIN

In the 1800s, many people were getting rich making dyes from coal tar. A German dye salesman named William Bayer teamed up with Johann Weskott, a professional dyer. They formed a company that made modern dyes using chemistry in an industrial way.

One of Bayer's most popular products was fuchsine. It was a powder made of green crystals, but when mixed with water, it produced a magenta-coloured dye like the colour of a fuchsia flower.

The industrial dye business had started with William Perkin, who had tried to make a medicine and ended up making a profitable dye. The Bayer company went in the other direction—after a few years making dyes, they started making profitable medicines.

One of their big successes was Aspirin or ASA. For thousands of years, the substance had been extracted from the bark of willow trees as a cure for aches, pains, and fevers. In 1897, the chemists at Bayer made a pure version of the chemical in the laboratory.

Aspirin would turn out to be a big success for Bayer, but the chemists' Bunsen burners had hardly cooled off before they were working on a second breakthrough painkiller. They wanted to solve the problem of morphine addiction. At that time, morphine was the most powerful painkiller available. Unfortunately, because it was made from opium poppies, morphine was highly addictive. The chemists thought it might be possible to make a similar drug in the laboratory—a painkiller just as effective as morphine but without the addictive side effects.

Less than two weeks after Bayer's scientists had produced Aspirin, they also managed to create their newer and safer version of morphine. The drug did an amazing job on aches and pains, not to mention suppressing coughs. It seemed to be even more powerful than morphine. They gave it the brand name Heroin, because they considered it a heroic drug—or perhaps because after taking it, they were filled with a strangely heroic feeling.

Bayer didn't make snake oil. They were a responsible pharmaceuticals company, selling their products through doctors and pharmacists. Medical professionals were thrilled by the new product, which seemed safe and effective. As with fentanyl today, Heroin required a doctor's prescription, and doctors told each other that although Heroin addiction was possible, it was rare.

The truth was that, although Heroin certainly relieved pain and suppressed coughing, Bayer's "non-addictive morphine substitute" was even more addictive than the morphine it replaced. When people took it, they had a euphoric feeling that helped get the addiction started.

It took a while for the company to realize the problem they'd created. In the meantime, Heroin was advertised alongside Aspirin. There was even a children's Heroin, to soothe the coughs of a child with bronchitis.

Bayer eventually stopped manufacturing Heroin in 1940, but by then, illicit uses of the drug had become unstoppable. Heroin, intended to be safe and non-addictive, remains one of the most addictive and deadly street drugs in the world.

THIS STINKS!

The race to extract new substances from coal tar was a scientific gold rush in the 1800s.

The invention involved in organic chemistry happens on a tiny scale. The carbon in coal can be rearranged like Lego bricks to form limitless combinations, many of which duplicate chemicals found in nature. As we've seen, though, the scientists didn't have a good understanding of the work they were doing, and most of the successes were accidental.

In 1888, an ambitious German chemist named Albert Baur made a different accidental discovery. He started out trying to make an extra-powerful version of TNT. He didn't have much luck. The chemical he made was a weak explosive. It was also a stinky one.

As he was airing out the lab from his latest experiment, it occurred to Baur that the foul stench seemed vaguely familiar. A keen-nosed expert might describe it as "revolting, a combination of earth, warm ripe sweat, and leather, with notes of spice and chocolate."

Like Perkin before him, Baur didn't dump the mixture and move on to the next experiment. He realized he might have invented something interesting. He recognized this odour as musk. And, long before Elon came along, the smell of musk was the smell of money.

That's because this substance, which smells vile in its concentrated form, can become very pleasant when it's diluted. The natural form of musk has long been used by the perfume industry. It was usually extracted from the anal glands of a certain Asian deer—a strange little animal with fangs instead of antlers. The male deer smears its hindquarters on branches to mark its territory. Hunters used to kill the deer just to cut out the glands, and each dead deer produced only around 20 grams of musk.

Baur patented his new artificial musk and also created a few other variations of the chemical. It was as successful as he'd dreamed. Just as the fabric industry had been excited by the new artificial dyes, the perfume industry embraced Baur's new synthetic scents. And if the deer had any sense of what was going on, they would have been grateful, too.

The synthetic musks Baur had invented were used to make some of the world's most popular perfumes, including the famous Chanel No. 5. (They have since been replaced by more advanced musks, which last longer and are safer.)

A flood of other synthetic musks followed, and some of those musk smells are now so familiar that we don't even recognize them as musks anymore. Modern synthetic musk chemicals are so cheap that they no longer need to be saved for expensive perfumes—they can be added to household products such as fabric softeners. The "fresh linen" scent, widely used in soaps and candles, is one example.

For Baur, the stink in his laboratory turned out to be the sweet smell of success.

NOT FADE AWAY

Sometimes the "mother of invention" isn't necessity but embarrassment.

James Morton was a Scot who owned a textile mill in the north of England. His company made attractive coloured fabrics using the new dyes that became available in the early 1900s. He was proud of his company's attractive designs and colours.

Around 1904, he took a trip to London. Some of his products were on display in a shop window—a set of specially designed tapestries. Morton himself had overseen the selection and blending of the colours. But he immediately saw that something was wrong with the way they looked. Some colours had faded, while others had changed hue. The end result looked terrible.

Morton assumed that, for the tapestries to fade the way they had, they must have been exposed to direct sunlight for a long time. He checked with the store owners. "How long have they been hanging there?" They said, "A week."

Morton was mortified. What good was it to make fabrics with attractive colours if exposing them to light for just a few days changed the colours beyond all recognition? He felt like a failure. His artistically designed fabrics were a disaster. He might as well just give up the business.

But after he had recovered from the shock of seeing how poorly his own goods performed in the real world, he started to think about the problem in a more systematic way. His company made cloth from materials that had been dyed by other companies, using both traditional dyes and the newer synthetic dyes. How did his cloth compare to products from other manufacturers?

He collected hundreds of beautiful fabrics from many companies, then hung them up in his greenhouse to see what would happen. Part of each sample was left exposed to sunlight and part was covered. The results were shocking—almost all the samples faded badly. In some cases, dark—and expensive—velvets became completely white within a week.

Like most manufacturers, Morton had never taken too much interest in the chemical details of the dyes used in fabrics. Now, humiliated, he realized what a mistake that was. The dyeing industry looked like it was about to be a dying industry.

Morton was determined to apply more quality control to dyes. In his tests, a few colours had held their colour in the sunlight. He vowed to test every dye, make notes of which ones held their colour, and try to build up a complete palette. He called the rare, durable dyes "sundour"—in Scottish English, dour means stubborn and hard to shift.

He wanted to put each colour through rigorous testing under bright sunlight. That was a problem because he lived in Cumbria. It was not quite as cloudy and dark as his native Scotland, but it's one of the cloudiest, wettest, rainiest parts of England. Fortunately for Morton, Britain still had an empire, and Morton's brother-in-law was an official in India. Morton sent the samples to him—strips of cardboard with various colours of thread wrapped around it. He included careful notes to identify each sample.

It took years of testing, but Morton eventually came up with a range of colours that were guaranteed not to fade in bright sunlight. There had been a few colours missing in his range, but Morton hired a chemist to develop "sundour" dyes to fill the gaps.

His dyes were expensive, and he wasn't sure if people would appreciate their unique quality. He needn't have worried. The public was getting tired of buying brightly coloured fabrics that rapidly faded, and they were willing to pay extra for colour-fast fabrics. Many big manufacturers, including Burberry, bought his materials to use in their products.

Morton's invention of fade-proof dyes work made him wealthy, won him the admiration of the public, and even earned him a knighthood.

An invention that came from professional embarrassment turned into an embarrassment of riches.

CHAPTER 17

GREAT NEW WAYS TO KILL

WAR often brings inventors out of the woodwork. Governments find themselves deluged with creative ideas for defeating the enemy.

The desperation that accompanies war sometimes means that officials will consider ideas from brilliant inventors they might have ignored in peacetime.

At other times, they may end up listening to crackpots.

A FLARE FOR COLOUR

In the 1840s, when Martha Hunt was a teenager of fifteen or sixteen, she eloped with a twenty-one-year-old named Benjamin Franklin Coston. Like his namesake, he was an inventor, and he had an important job running a scientific laboratory for the US Navy in Washington.

Unfortunately, the husband's work exposed him to dangerous chemicals. The toxins made him increasingly sick. He eventually died.

At the age of twenty-one, Martha became a widow.

Going through her husband's papers, she reviewed some of his ideas. One particularly impressed her—it was an innovative naval signalling system. In the nineteenth century, ships usually communicated by raising combinations of flags, which could spell out brief messages. That worked in the daytime, but it was difficult to communicate at night. Her late husband's idea had been to replace the flags with coloured flares.

Martha Coston developed the idea, designing a system that was both simple and reliable. When she started, she may have been trying to honour her husband's memory, but as she worked on the problem, she had to become an inventor herself. It took her years to find the right chemicals for the flares. She realized the colours needed to be easily distinguished from each other, and each colour of flare also needed to have a similar brightness, so signals could be read reliably over a long distance.

Coston wanted flares in three colours, and hoped for red, white, and blue, to match the American flag. She eventually created good flares for red and white, but had difficulty creating a suitably vivid blue, so she went with red, white, and green instead—less patriotic, but perfect if you want to give your naval battle a Christmas vibe.

The flares were packaged as cartridges, launched from a specialized pistol. Some flares fired a single colour, while others burned a sequence of two or three colours, like red-white-red. Colours were clearly marked on the outside of each cartridge. Combinations of flares could send complex coded messages. The system worked well, and the navy bought thousands of them. They also agreed to pay Coston a large fee for the rights to her patent, but they were slow to pay and only gave her half of what she had been promised.

When the American Civil War began, Coston knew the navy would need her flare system. Motivated more by patriotism than profit, she agreed to supply flares to the Union at cost. This turned out to be a bad move. Her official "cost" had been based on the price of her supplies during peacetime, but when the country was at war, everything became more expensive. Coston was still obliged to sell at the agreed price, and the more flares she sold, the more money she lost. She spent years trying to get her money back from the government, but she managed to keep her business going and turn it into a financial success.

The flare system Martha Coston invented was eventually adopted all over the world, and similar flares are still used today. They have signalled emergencies, disasters, calls for help, and victories—Martha Coston could have used them to write a biography.

GUNCOTTON APRON

Guncotton was a breakthrough in explosives and led to the replacement of gunpowder in guns and artillery shells. As a weapon, this explosive has been the cause of many tragedies, but its accidental discovery plays out like a sitcom.

It happened in 1845. A Swiss chemistry professor named Schönbein loved doing experiments. His wife told him that if he wanted to work on his chemistry experiments, he should do it in the lab at the university—under no circumstances was he to do it at home. But later, she went out, and when she was gone, Schönbein snuck down to the kitchen to do some sneaky chemistry.

He was having a fine time mixing chemicals together, and he was busy mixing two common acids—nitric acid and sulfuric acid—when he had an accident. The container spilled, and the acid mixture splashed all over the kitchen.

He didn't want his wife to discover that he had been up to his old chemistry hijinks, so he hurriedly grabbed a nearby cloth and used it to mop up the spilled acid.

Then he realized that the cloth he'd used was his wife's apron. He carefully hung it up in front of the stove to dry before she got home.

(He didn't seem to consider that his wife might return early and wrap herself in an acid-soaked apron. Or perhaps that was part of his plan. Either way, it's not what happened.)

As the acid soaked into the cotton, it caused a chemical change. As he watched the drying apron, he suddenly saw a bright flash of flame. A fraction of a second later, the apron was gone, entirely burned up. He had invented guncotton, which could produce a spontaneous explosion—

although it was probably no match for the explosion he faced when his wife saw what he'd done to her apron.

x x x

Schönbein had made an important discovery. The guns and cannons of the time all used gunpowder, which makes an impressive bang, but also produces copious quantities of smoke. Guncotton also provides an explosion, but with just the faintest wisp of blue smoke. Armies could say goodbye to the "fog of war." Guncotton was also more powerful than gunpowder—guns could fire a bullet further and at higher speed.

Once word got out, people were impressed by the new discovery, and factories were set up to produce it.

Unfortunately, it wasn't quite the lottery win that Schönbein had hoped for. Guncotton was more powerful than gunpowder, but it was so much more powerful that it tended to destroy the barrel of a gun, so the military didn't end up adopting it. Its instability was also a problem—guns could easily go off as they were being loaded, and the factories where guncotton was made had a bad habit of exploding.

It took decades for other chemists to solve these problems. Guncotton eventually morphed into cordite, which became an important explosive in firearms.

Something closer to the original guncotton recipe is still used by magicians as flash paper, making small objects seem to "disappear in a flash." Schönbein could have told them that it also works for aprons and marriages.

MACHINE GUNS FOR PEACE

American inventor Jacob Perkins came up with a number of innovative ideas in the early 1800s. He built steam engines, a printing press for currency, and one of the first refrigerators. But his most spectacular and controversial invention was a remarkable gun. The gun was enormously powerful, it could fire continuously, and it didn't need gunpowder. Its explosive force came from high-pressure steam.

The Duke of Wellington, the same austere British officer who had defeated Napoleon, was intrigued by what he had heard about the new weapon. Thanks to his support, the gun was tested in England before a group of government officials and high-ranking military officers.

The weapon was aimed at planks of wood. The tests started with single shots. As a rule, a bullet that penetrates 6 millimetres into a pine plank will also inflict serious wounds on a human. Firing from a distance of 32 metres, a bullet shot from Perkins's steam gun went straight through eleven 25-millimetre planks.

The first plank tests involved just one bullet, but the weapon's large supply of steam allowed multiple shots. Perkins switched barrels and added a tube holding a large stack of musket balls. The projectiles dropped through the tube into the barrel and were blasted out with incredible power. The weapon was a steam-powered machine gun that could shoot up to a thousand rounds a minute.

As the amazed officials looked on, the gun was tested against armour plate. The lead bullets struck the metal with such force that they were flattened. When lead musket balls were switched for iron balls, they easily punched a hole through 6 millimetres of armour. It was an impressive display.

In 1824, an editorial in the *London Mechanics' Register* discussed the new gun in awed tones, presenting an argument eerily similar to the discussion surrounding nuclear weapons—after describing the weapon's power, the editor assumes that most readers will say "What a murderous instrument! What barbarians to recommend such a mode of warfare!" But he went on to say that this nineteenth century "weapon of mass destruction" was so horrifying and deadly that it might end all war, because if it were ever to be used, the rate soldiers were slaughtered would be far greater than the rate they could be replaced.

> What plague, what pestilence would exceed, in its effects, those of the steam gun?—500 balls fired every minute . . . one out of twenty to reach its mark—why, ten such guns would destroy 150,000 daily. If we did not feel that this mode of warfare would end in producing peace, we should be far from recommending it.

Not only could a single steam gun match the performance of a hundred soldiers with muskets, but it was also much cheaper to fire. Gunpowder was expensive—shooting 15,000 bullets cost £525. Firing the same number of bullets with steam power cost only £4 in coal.

Britain was not the only country looking at Perkins's invention. France was also interested. It was even rumoured that the czar of Russia wanted to buy the weapon, although so far Perkins had been too patriotic to consider it. (This sounds suspiciously like a rumour that might have been started by Perkins.)

But after the tests were over and government officials had thought about it for a while, they told him no. They would rather just stick to old-fashioned gunpowder for their firearms. Perkins arranged for tests with the French army. They didn't buy it either. He didn't manage to sell his steam gun to anyone.

It sounds like a case of stuffy bureaucrats who lacked the imagination to support a brilliant new invention. But the bureaucrats weren't fools—they had some good reasons for rejecting the invention. Although Perkins's steam gun did everything it claimed, it had some big disadvantages. It required a steam boiler and that was heavy—the whole setup weighed around 5 tons (about the same as seventy soldiers). It would be a cumbersome piece of equipment to set up. And while gunpowder might be more expensive than water, it was ready at a moment's notice, whereas Perkins's gun couldn't be fired at all until the boiler had been brought up to the correct temperature.

Although the gun was never sold to the military, it did end up making money for Perkins. When Perkins was older, he set up a museum in London, showcasing his many inventions. People paid a shilling for admission, and the most popular exhibit was the steam gun, which, every day, would demonstrate its power by blasting a volley of shots down the shooting gallery.

Perkins had faced some frustrations in his inventing career, but at least he found a good way to let off steam.

CART BEFORE HORSE

Inventors must often "think different" and defy popular wisdom.

For Serge Berditschewsky Apostoloff, an inventor living in London around 1905, that meant defying the proverb "Don't put the cart before the horse." His invention was simply a cart with a horse harnessed from behind. The horse was partly boxed in, pushing blindly and obediently, while the driver used a regular steering wheel fixed to the front wheels.

What most excited Apostoloff was that, with the horse out of the way, he could give the driver a large forward-mounted gun, "so that a constant fire can be maintained, for example during the pursuit of a retreating enemy."

Not so good, of course, if you're the one retreating and you need to shoot backwards. Still, the idea of a large forward-mounted gun on a vehicle would probably appeal to many modern drivers.

MAYNARD CAPS

In the 1800s, Edward Maynard was one of the most prominent dentists in America, with a clientele including many of the leading politicians of the day. It's even said that he was offered the position of court dentist to Russia's Czar Nicholas I.

He did extractions and root canals, but his biggest success was with caps—not the caps that go on teeth, but the ones that go in firearms.

Maynard was as interested in weapons as he was in dentistry, and his mind went back and forth between gums and guns.

In Maynard's day, many soldiers and hunters were still shooting with flintlock muskets. The weapon used a spring-loaded mechanism. When you pulled the trigger, a piece of flint slammed down onto a metal plate to create a spark. The spark lit the gunpowder in a depression at the top of the gun, then the flame burned and fizzed its way through a little hole into the barrel of the gun, causing the gunpowder inside to explode. This series of events meant there was a brief but noticeable delay between pulling the trigger and firing the gun. Sometimes the delay was long enough for a bird or animal to hear the click of the trigger and move out of the way before the bullet fired.

One improvement that some people used was to ditch the flint and use a percussion cap—a tiny cylindrical cup made of copper. The cap contained a small quantity of mercury fulminate, a chemical that would explode when struck. If the hammer of a gun struck a percussion cap, it would go off almost immediately, igniting the gunpowder inside the barrel. It solved the speed problem. That is, until it was time to reload. The caps were small and fiddly, and you had to carefully insert a new one onto the gun each time you fired a shot. It was difficult to change the percussion cap quickly if a soldier was wearing gloves, or if he had cold hands, or if he was flustered because someone was running at him with a bayonet.

Edward Maynard's bright idea was to replace the troublesome metal percussion caps with a strip of "tape primer"—a paper roll with a small amount of explosive sandwiched inside it at intervals. This was basically the same as the roll of caps used in toy guns. When you pulled the trigger, the gun fired, and a mechanism pulled the next blob of explosive into place.

You still had to reload your gun with gunpowder, but there was no more fiddling with those nasty little percussion caps. Maynard used his political connections to demonstrate his design to the military. They were impressed and carried out tests. The tape worked well, and they placed an order for a firearm that would use the new system—the Springfield Model 1855 musket—a firearm with a hole on the side to hold a roll of paper caps. It looks incongruous today, but in the American Civil War, cap guns were deadly weapons.

Unfortunately, Maynard's system didn't always work reliably in wartime conditions. The mechanism that advanced the caps was delicate—easy to break and difficult to repair. And although the paper tape was coated with wax to make it waterproof, it often failed to fire in wet weather. Eventually, the military gave up on caps of all kinds and instead switched to modern-style cartridges, which have the shot, gunpowder, and detonator all sealed in a brass cylinder.

Maynard came up with many other gun-related inventions over his life, but the military use of his inventions was short-lived.

Maynard might have made more money from his innovations if he'd considered other uses for his "tape primer." When the military orders dried up, the companies that had manufactured real guns for soldiers kept the same two percussion cap systems and used them in toy guns for children. The percussion caps were changed from metal to plastic and today are often sold in rings. The paper caps remained similar to Maynard's original invention and would be used in millions of toy guns over the next hundred years.

A BOMBSHELL'S TORPEDO

During World War II, Hedy Lamarr was considered one of the most attractive female stars in Hollywood, a "bombshell." But this famous actress also had a keen mind. When she wasn't acting, one of her hobbies was inventing, and the bombshell favoured another kind of explosive.

In the 1930s, while she was still a teenager, she had married an arms manufacturer named Friedrich Mandl. They were not very compatible, and she eventually left him, but during their time together, she learned a thing or two about weapons systems, including ideas for a remote-controlled torpedo.

Torpedoes were powerful, but not very accurate, especially when they had to travel long distances. After a torpedo had been fired from a ship or submarine, it could easily get caught in ocean currents and stray from its course. That wasn't the only problem. Even if the torpedo was aimed correctly and travelled straight, a ship might still see the torpedo coming and change direction to avoid it. But if the person firing the torpedo could watch its progress and control it using a radio signal, its path could be continually adjusted until it hit the enemy ship.

But adding radio-control to a torpedo introduced a host of new problems. An enemy could listen for the radio transmissions and jam the frequency. They might even send signals of their own and turn the torpedo back at the vessel that fired it.

It wasn't enough to have a radio-controlled torpedo. To make it a good weapon, the torpedo also needed a secure radio system that was immune to interference—which was a difficult problem.

Lamarr couldn't see a solution until she talked with one of her Hollywood neighbours, a composer and inventor named George Antheil. He had written a musical work named Ballet Mécanique. It was a kind of robotic symphony, which included sixteen player pianos, synchronized to play together. The pianos used a paper roll, punched with holes. As the holes passed through the mechanism, the corresponding keys would play.

Antheil's work was exciting to Lamarr because it seemed to provide a solution to the torpedo problem. The same mechanism that played eighty-eight notes on synchronized player pianos could instead change the frequency on synchronized radios, jumping among eighty-eight frequencies. If two radios were synchronized this way, with each one changing frequencies at the same time, they could stay connected to each other, but anyone trying to listen in without a copy of the synchronizing device would only hear chopped-up nonsense signals.

The method wasn't limited to torpedoes. It could also be used for secret radio communications between military units, or to control "aerial torpedoes"—which we'd now call drones.

Hedy Lamarr and George Antheil worked together on the new secure radio signal and filed a patent for it in 1942. They pitched the idea to the navy, but officials said their system would take up too much space inside a torpedo. The navy recognized the possibilities, though, and classified the patent "top secret." The inventors didn't get any credit on their idea for years.

There were likely a few reasons that the military wasn't interested in Lamarr's ideas. Even if Lamarr and Antheil had reduced the size of the device, it would have been a challenge to keep the delicate mechanisms synchronized—especially if one was on a submarine and the other was being blasted into the ocean. Antheil had faced enough problems keeping his pianos synchronized—and that was on dry land. In his concerts, he ended up having to use one player piano and a group of human musicians.

Another problem was that Lamarr's information was out of date. She had heard about radio-controlled torpedoes in the 1930s, but the military had since lost interest in the idea. Their focus had turned instead to the "homing torpedo," which listened for the sound of a ship and aimed itself in that direction.

Lamarr's device could have been used for other radio communication, but unfortunately her idea wasn't that new. Radio operators had been exploring frequency hopping for years, and a 1926 patent by Otto B. Blackwell describes a "secrecy communication system" that used paper rolls with holes. His version was based on the punched tape used by telegraph operators, but the idea was very similar.

Perhaps the final reason the US military didn't jump at Lamarr's invention was that they already had one. They had spent a fortune on a top-secret communications device known as SIGSALY. It scrambled voice communications in a random way. As with the other systems, both the sender and receiver needed a synchronized signal. Lamarr's had two player-piano rolls. SIGSALY used a pair of identical phonograph records, which allowed much faster scrambling than a punched tape. The disks at each end needed to start at exactly the right time and spin at exactly the same speed. The machinery was huge, clunky, and very expensive, but it worked, allowing thousands of secure phone conversations between leaders and commanders in Britain, North America, and even the South Pacific. After the war, it was discovered the Germans had listened in on the signals but hadn't been able to make anything of them—they couldn't even tell that they were phone conversations.

Ironically, one of the people behind SIGSALY was another person who didn't get much credit for his inventions during his lifetime—the British mathematician Alan Turing. He was long dead before his many contributions were made public.

Lamarr did finally receive some credit while she was alive. She had always resented being admired for her beauty rather than her brains, but in the 1980s, when "frequency hopping" was being widely used for technologies like cell phones, engineers discovered her old patent, and she received wide recognition for her invention. While her own piano-roll approach may not have changed the world, she was one of a few inventors who were thinking along the same lines, and today "frequency hopping" is an important principle in digital communication.

TWIRLIN' THE MERLIN

Until the 1960s, mechanical engineering was considered a man's field, and those women who managed to gain a foothold were often unusual personalities with exceptional skills.

That certainly applied to Beatrice "Tilly" Shilling. She'd been fascinated by engineering since she was a child. She made herself an expert on electronics, on engines, and on practical shop skills like brazing. By the 1930s, she had graduated as a professional engineer.

She worked on motorcycles and was skilled at racing them. The male racers didn't take her seriously at first, as she walked around in her heavy coat and Harry Potter glasses. But when she climbed on the custom bike she'd modified herself, she left them in the dust. It was said that when a fellow engineer wanted to marry her, she refused unless he could run the Brooklands motorcycle circuit at 160 kilometres per hour, as she had done years earlier. He eventually succeeded, and she married him.

Even in an age where women were expected to take on "helper" jobs, the military recognized Shilling's formidable skills, and put her to work in the research division of the Royal Air Force to work on engine carburetors—the chamber inside the engine where fuel and air mix together. They needed help from the best because they had a very big, very deadly problem.

Britain's finest fighter plane was the Spitfire. It had evolved from a series of racing planes, and was very fast, thanks to its powerful Rolls-Royce Merlin engine. Its main German opponent was the Messerschmitt Bf 109—it was smaller than the Spitfire, and not quite so fast, but it was nimble in the sky. Not only could it manage a tighter turn than the Spitfire, it also had a useful trick up its sleeve—thanks to its fuel-injected engine, which lacked a carburetor, it could fly upside down.

The Spitfire couldn't do that. Neither could its sister plane, the Hurricane, which also used the Merlin engine. Despite the engine's power, the Merlin relied on gravity for the fuel to flow correctly into the carburetor. A Spitfire could manage a quick roll, but if you flew a Spitfire upside down for more than a moment, the carburetor flooded with fuel, the engine stalled, and the plane might crash.

Aside from impressing audiences at an airshow, this upside-down trick may not sound much of an advantage for the Messerschmitt, but German pilots soon figured out how to exploit it. They realized that the

Spitfire's limitation also applied to power dives. When a plane turns its nose down to enter a dive, everything inside the plane rises rather than falls—it's just like being upside down.

The Germans started using this trick to gain the upper hand over the Spitfires. When a Messerschmitt was being chased by a Spitfire, the German pilot took his plane into a power dive. The Messerschmitt's engine handled the reversed gravity just fine, but if the Spitfire tried to follow, its own fuel would flow the wrong way into that pesky carburetor, and it would stall.

There were some countermeasures the British pilots could take to chase the German planes safely—ironically, it involved briefly rolling the Spitfire upside down, then diving before the engine stalled—but this bit of fancy flying took precious seconds, and once the manoeuvre was complete, the enemy fighters had often escaped. Pilots complained that they needed a solution.

The people in Tilly Shilling's department had tried to fix the problem with an improved carburetor. It failed in testing. Unfortunately, the only solution the engineers could think of was to redesign the whole engine—but that was a huge job and could take years.

That's when Tilly Shilling stepped in with her solution. She knew it would be impossible to make the plane fly upside down for an extended period, but the manoeuvres causing all the problems only lasted a few seconds. If the engine could be tweaked so it could run upside down for that long, it might be enough to make the difference. And she knew just how to do it.

Her idea was to put a metal stopper in the fuel line, with a tiny hole in the middle. She used a brass thimble at first. The hole would let through exactly enough fuel to run the engine at full power, but because it restricted the flow of fuel, the carburetor wouldn't have a flooding problem. In a dogfight, this simple "cheat" allowed the Spitfire pilot to match the Messerschmitt's movements.

It was such an easy modification that any mechanic could quickly install it on a plane. The pilots were delighted with the improvement and deeply grateful to the woman who had designed it—although that didn't stop them naming the part "Miss Shilling's Orifice."

Shilling stayed in her job until she retired, receiving honours from the government. She went on to apply her engineering expertise to everything from bobsleds to Formula 1 racing cars to intercontinental ballistic missiles.

CHAPTER 18

AS SEEN ON TV

MANY a clever idea has failed because it didn't have enough marketing push behind it. The opposite is also true—some new ideas that weren't especially brilliant have been turned into bestsellers through clever marketing.

If you watched TV in the 1960s and 1970s, you couldn't miss ads from K-Tel. Most were for compilation record albums ("Original hits! Original stars!"), but some were for innovative products, available only by phoning the toll-free number at the end of the ad, where "operators are standing by."

K-Tel was a Canadian company started by Philip Kives. He was born in the blink-and-you've-missed-it hamlet of Oungre, Saskatchewan, about 150 kilometres south of Regina.

His ads played often and were annoying but effective: a fast-talking announcer described the features of a product, accompanied by actors showing their frustration with the old way and their delight with the new.

Many of the early products sold by K-Tel were inventions of an inventor named Sam Popeil, who invented the Dial-o-Matic (grater) and Veg-o-Matic (slicer and dicer). His son Ron Popeil then started his own company, Ronco, which ran ads with a similar style to K-Tel's.

Later TV hucksters like Billy Mays and Vince Offer continued the trend, often selling the same old products under a new label.

GINSU KNIFE

The Ginsu knife is one of the best known "as seen on TV" products. On commercials, it was shown cutting through metal, frozen food, and wood, then making a neat cut through a tomato. It was sold with an array of smaller Ginsu knives.

The Ginsu used an inexpensive stainless-steel blade, which was given a serrated edge in the factory using two rows of rotating wheels. This method of manufacture was invented in New York in 1919 and used to make the first serrated bread knives.

The Clyde Castings Company of Ohio were specialists in making these inexpensive serrated blades, using various serration patterns. They had sold knives for decades under the name Quikut. Their products were often given away as part of "free steak knives" offers.

For the TV ads, the sellers replaced the Quikut brand name with something that sounded Japanese. At the time, North America was both interested in and threatened by Japanese culture and technology.

Although cheap to manufacture, the knife was indeed very sharp and surprisingly durable, but when it finally did go blunt, it was impossible to sharpen. In marketing speak, that translates to "Never needs sharpening!"

SHAMWOW AND ZORBEEZ

Aggressively marketed by TV pitchman Vince Offer, ShamWow is a synthetic cloth. On TV ads, it competed with a similar product, Zorbeez, sold by Billy Mays.

Real chamois (or "shammy") is made from leather. It originally came from the chamois goat, but sheep leather can produce a similar material. A chamois is highly absorbent and doesn't leave lint when it wipes a surface.

In the 1960s and 70s, scientists found ways to get a similar performance from synthetic fabrics. This achievement involved a series of fabric breakthroughs in different parts of the world to produce finer and finer filaments—after spinning ultrafine synthetic fibres from plastics like polyester and nylon, the fibres were made even finer by splitting them along their length.

When fabric is made from these "microfibres," it has some useful properties. It insulates well. The tiny fibres have an electrostatic charge that can trap dirt particles. The enormous surface area of the fibres also allows the cloth to absorb large amounts of liquid. The finer the fibres, the more absorbent the cloth.

The TV products are made from these remarkable fibres—they are inexpensive microfibre cloths, cleverly pitched and sold at a huge markup. Order today!

SLAP CHOP

The Slap Chop was a cylindrical chopper. Press down on the plunger and a set of blades chops whatever is inside—usually vegetables—then rises again. The blades turn slightly on each stroke, making it easy to chop something into small pieces.

The Slap Chop was first sold in 2008, but it was just one in a series of plunger-style choppers. The Chop-o-Matic from 1956 was among K-Tel's first offerings, using a product patented by Sam Popeil.

The basic idea goes even further back. A very similar chopper, designed to slice onions, was patented by an inventor named John Edwards, living in Bushire, Persia (now Bushehr in Iran). That was in 1891.

As exciting "new" products go, this was one of the oldest.

BUT WAIT, THERE'S MORE

The "as seen on TV" products included dozens of odd inventions. Some are still sold, while others are now just memories.

BOTTLE CUTTER KIT

"Save those empty bottles!" cries the announcer. Rather than throwing bottles away, they could be turned into a glass tumbler, or vase, or a unique ashtray. The bottle cutter kit consisted of a roller that held a bottle while it was rotated and scored by a glass cutter. The actual cutting process posed more of a challenge. The instructions told the user to rotate the scored bottle over a candle, then rub ice on it to separate the glass. If it separated cleanly, the user would have to sand down the sharp edges to make the glass safe to use. There are other hitches. First, the user would have to figure out where to put all those sharp bottlenecks. And after recycling bottles for a few months, they would need to figure out how to dispose of hundreds of unwanted tumblers, vases, and unique ashtrays.

DENTAL MATE

"For clean, healthy teeth, you should brush after every meal. Now you can!" says the ad's announcer. It's not clear what had been preventing people from brushing their teeth, but Dental Mate was there to fix it. The product was a toothbrush with a built-in toothpaste dispenser. Just press on a cartridge in the handle to dispense toothpaste directly into the brush's bristles. Although "toothpaste" doesn't quite describe the thin, mint-flavoured liquid exuded by the brush. Dental Mate came with the toothbrush and a single cartridge of its strange, runny tooth-liquid. Good luck finding refills.

POPEIL'S POCKET FISHERMAN

This compact all-in-one fishing rod was small enough to go into a glove compartment or a pants pocket—assuming your pants had enormous pockets. The handle contained a hook and float. A tiny rod folded out from below. But what problem was Pocket Fisherman designed to solve? How often do people encounter the kind of fishing emergencies that require a miniature plastic fishing rod? Owners say it will catch fish, but so will a length of cotton and a bent needle.

FISHIN' MAGICIAN

This went up against Pocket Fisherman in the battle of the TV fishing gadgets. It had "eleven fishing aids" including a built-in tape measure, weighing scale, knife, and bottle opener. Basically a "fish army knife."

K-TEL RECORD SELECTOR

This was a rack to store LPs. The records each slotted into a plastic clip. When one record was flipped forward, it would pull the next one, and then the next. The gizmo had some limitations: it didn't work with double albums, and the plastic clips had a bad habit of damaging the bottoms of the album cover. The device was offered as an easy method to review your record collection. An even easier method was to stack the records sideways and read the print on the spines.

BIGFOOT

Bigfoot was a pair of oversized, foot-shaped soles that children could slip over their shoes. They could make huge footprints in the snow and then look at those footprints. It was hours of fun. Okay, perhaps minutes of fun. Or seconds, certainly.

BETTY ANN'S TAILOR-MARKER

Also sold as Aunt Mary's Tailor-Marker, this was a lightweight sewing awl that pushed thread through a piece of fabric to mark it with a series of rapid stitches. It seemed like it would be a fast way to repair clothes, but each push of the needle only made loops of thread on the other side of the fabric—the stitches weren't held in place, and the lightest tug would pull them all out again. Marking was the only thing it was good for, and for many jobs, chalk or pins would work just as well.

BEDAZZLER

One of the more successful "as seen on TV" products, the Bedazzler was a plastic device like a stapler. It stamped spiked studs and rhinestones onto clothes or fabric. The advertisement showed belts and bags being decorated, while jackets were transformed into suits of glittering rhinestone armour that would light up any room, and slice up any couch.

VAQUA-DAPT

Vaqua-dapt was a bucket that attached to any home canister vacuum cleaner. A second hose ran from the bucket to a vacuum nozzle. The container basically turned the vacuum cleaner into a shop vacuum, with water or solid objects going into the bucket rather than into the vacuum cleaner.

BRUSH-O-MATIC

This brush from K-Tel used a velvet-like surface to pick up lint. Many K-Tel products were the something-o-matic. TV competitor Ronco preferred the word "miracle," so their version of this brush was named the "Miracle Brush." Because the fibres on the brush all point in the same direction, they will pick up dust with a forward stroke and release it if wiped on cloth in the opposite direction.

This was a popular "as seen on TV" product and worked quite well. Similar items are now common in regular stores. The underlying invention seems to have come from Switzerland in the mid-1960s, when researchers found an inexpensive way of sticking very tiny synthetic bristles to a fabric backing so they all pointed in the same direction.

THE PATTI-STACKER

The Patti-Chef or Patti-Stacker was a plastic cylinder with a plunger. Put a chunk of ground beef inside and the plunger will squeeze it into an even circle. Plastic disks dropped in after each chunk of beef allowed the user to make a stack of hamburger patties, although the user would have to press each new patty down separately, so there seemed to be no time saved by putting them in a stack.

K-TEL BODY BUILDER

This was a pair of plastic handles with an angled edge that fit the side of a table. Any fitness buff and table enthusiast could grip the handles and do push-ups against the table. A person could probably do the same thing against a wall without the handles—or just do push-ups on the floor.

SCHTICKY LINT ROLLER

Another Vince Offer offering, this fascinating reusable sticky roller used a silicone-based plastic to pick up dust and lint. Although the roller's stickiness runs out quickly compared to an adhesive-based lint roller, it can be made sticky again if it's washed under water and left to dry.

The technology behind this product is based on silicone polymers, which were first used around World War II as a synthetic replacement for rubber. By changing the amounts of certain chemicals in the mix, the silicone plastic can be made more elastic and tacky. The same idea was the basis for some sticky toys in the 1980s, including a short-lived fad of "wall walkers"—silicone creatures which seemed to crawl down walls on long, sticky legs.

CHAPTER 19

BURNING BATS AND SUICIDE PIGEONS

THE military forces of the world have often used animals in warfare. The most obvious examples are horses, which carried knights and cavalries into battle, but many other animals have been conscripted for wartime duty—elephants, mules, oxen, and, in recent years, even beluga whales.

However, some ideas that some people have dreamed up for turning animals into weapons are not only inhumane, they're just plain weird.

LET SLIP THE SLUGS OF WAR!

Those slimy slugs chewing on your garden plants might not seem an obvious ally during wartime, but one inventor tried to use them that way.

Dr. Paul Bartsch was an expert on molluscs in the early 1900s. One day, a few of the slugs he'd been studying managed to escape their enclosure in his home. The scientist found them later on in the furnace room, but they were behaving strangely, contracting their bodies in a way that suggested they were in distress.

He realized that they were responding to the gases coming from the furnace. He found this fascinating and took notes on his observations—although maybe a better plan would have been to open a window and get the furnace serviced.

Some years later, during World War I, soldiers in the trenches were subjected to gas attacks. When an attack started, the soldiers only had seconds to get their gas masks on.

Bartsch remembered that the slugs in his furnace room had been uncomfortable in air that seemed fine to his human nose. It gave him an idea for an invention—a slug-based early warning device for poisonous gas.

He carried out tests on some unlucky slugs and discovered that, sure enough, the soft, wet molluscs were far more sensitive to poison gases than humans. Better still, the slug closed its breathing hole when the air turned bad, so you could gas a slug a few times, and it would usually recover afterwards. If soldiers were issued with gas-detecting slugs, they could put on their gas masks the moment the slugs gave the signal, long before the gas reached levels that were dangerous to humans.

In 1918, Bartsch wrote up a report on the idea and sent it to the US Army.

It's been said that the early warnings from this "slug brigade" saved thousands of lives. In the absence of dates and locations, we're a little skeptical about that. Unlike the canary in the coal mine, the slug doesn't make any sounds, so using this early warning system would have meant soldiers spending their time squinting at their regimental slug farm, watching for odd behaviour.

It would certainly have worked in theory. The slugs could have provided an early warning of a gas attack, but it seems like a better one would have been the sound of an exploding shell, and the sight of an approaching cloud of yellow-brown gas.

So, did Bartsch's slug invention save the lives of thousands of soldiers? We're taking that claim with a pinch of salt—which is another thing the slugs wouldn't like.

BUOYS AND GULLS

While armies were trying to harness the fighting powers of the humble slug, navies looked at ways to employ the majestic seagull.

In World War I, German U-boats were sinking many Allied ships. The public in Europe and America were full of ideas for dealing with the new undersea menace. A farmer suggested strapping hay bales to the sides of ships to protect them from torpedoes. Another writer proposed setting up strong magnets to catch and hold the U-boats. One man suggested putting barrels of Eno's Fruit Salts (a popular antacid) at the bottom of the sea—they could be opened by remote control, and the rising bubbles would (he imagined) lift the submarine to the surface.

One weird suggestion came from a man named Frederick Inglefield. His anti-submarine invention involved using flocks of wild seagulls as a submarine-spotting force.

Inglefield figured the seagulls could be easily trained for military service. A submarine—or a periscope on a raft—would travel up and down the English Channel, periodically releasing food. The seagulls would spot the food and eat it. After this routine had been repeated for a while, the hungry seagulls would learn that a periscope meant food. From then on, the birds would greedily assemble whenever their sharp eyes spotted a submarine. Human observers would no longer need to spot a tiny periscope sticking above the waves. They would only need to watch for unusual congregations of seagulls to know that a submarine was on the prowl.

Inglefield's plan didn't stop with just spotting the submarines. He wanted the navy to train seagulls in the same way the army trained pigeons. If the birds could be persuaded to poop on a sub's periscope, the enemy submarine would be blinded and would need to surface to give the periscope a wipe-down, and that's when they could be attacked by warships. Inglefield was vague on the details of how this seagull poop training would work.

There were plenty of crackpot inventions during the war, and Inglefield's plan sounds like one of them, but he had a few advantages over the competition—particularly in the fact that they were just armchair inventors, and he was a senior British admiral.

He was in charge of a group of motorboats whose work included locating and destroying submarines. Most of his boats were unarmed, which seems like a serious impediment but was no barrier to the admiral's inventiveness. Crews were told that, if they spotted a periscope, they were to creep up on it in their boats. Two swimmers, one equipped with a black bag and the other carrying a hammer, would swim to the periscope. One man would quickly pop the bag over the periscope, temporarily blinding the submarine, and the other would hit the periscope again and again with his hammer, breaking the glass lens.

The scheme had not been successful in sinking or damaging any submarines. It was difficult for the boats to be in the right place at the right time. But why use swimmers to sabotage a periscope when the same job could be carried out by seabirds?

So, the navy put Inglefield's ideas to the test, building a set of fake periscopes, which they towed around the waters of Poole harbour in Dorset.

The teams did their best to make seabird warfare work. Their fake periscopes spewed out food, and the seagulls flew in to eat it, but they didn't stick around when the food was gone. There were other issues too—submarines often operate in open ocean, but seagulls prefer the coast. In the end, no German submarines were spotted or strategically pooped on by Inglefield's seabird squadrons.

The whole idea was dumber than a bag and hammers.

LIKE A BAT OUT OF HELL

Some dentists have killer instincts. We've described how the dentist Edward Maynard invented a new kind of rifle in the American Civil War. In the following century, during World War II, another dentist also came up with a different and even more unusual weapon system.

Dr. Lytle S. Adams ran a dental practice in Pennsylvania. When he wasn't fixing teeth, he spent his time coming up with new inventions to change the world. He had tried to start a non-stop airmail service for the country, with biplanes picking up bags of mail by hooking onto them as they flew by, in the same way that trains could hook mail bags from poles. His plan didn't work out, but Adams remained optimistic and looked for new inventing opportunities.

In December 1941, Adams visited Carlsbad Caverns in New Mexico and was amazed at the sight of millions of bats emerging from the caves in the evening. But on his drive home, he heard terrible news on the radio: the Japanese had attacked Pearl Harbour, and America was now at war. Putting the two experiences together, he immediately thought of a new flight-based invention that could help the war effort—he would conscript those bats to work for the United States military.

His plan was to take a large number of bats, strap tiny incendiary bombs to each one, then release them over Japan. The bats would fly down and look for a place to sleep—following their instincts, they would crawl under the eaves and into the roofs of buildings. After the bats were settled down for the night, their bombs would trigger, setting off a fire in each building. With all the buildings simultaneously ablaze, the fire-fighting services would be overwhelmed, and the Japanese cities would be destroyed.

He wrote out his idea and sent a letter to President Roosevelt. If he had gone through the usual channels, this eccentric plan might not have received any attention, but Adams happened to be a friend of Eleanor Roosevelt. The president thought the idea seemed a little "batty" but conceded that there might be something to it. After checking with some government biologists, Roosevelt gave it the go-ahead.

Adams did express some qualms about burning Japanese civilians, but he had none at all about the cruelty of burning millions of bats. Some people may marvel at the bat's flying ability and its sonar hearing, but in Adams's view, bats were "the lowest form of animal life." In fact, he couldn't understand why God had created them, but he now figured it was probably all for this pivotal moment, to help America win the war by torching an enemy city.

The blazing bat-tallion was to be overseen by the air force, but Adams was allowed to run the project. The plan came to be known as Project X-Ray. He assembled a small team. Two members were the sorts of people you would expect—a scientist who studied mammals and an explosives expert. The rest were a collection of oddballs, including a former lobster fisherman from Maine, a hotel manager, a bodybuilder, a former gangster, and an actor—they seemed less like a research group than a crack heist team.

Adams and his team tested the capabilities of various bat species and eventually selected the Mexican free-tailed bat as the perfect candidate. It's a medium-sized bat, easily found in the US. Of all bats, it can fly the fastest and highest. These bats were also strong flyers, and the team discovered that a single bat could carry slightly more than its own body weight, allowing the transport of a miniature 15-gram incendiary bomb.

They built a bat carrier, which looked like a bomb with air holes. Each carrier contained over a thousand bats on steel trays. Each bat had an incendiary device glued to its chest. The tiny bombs had their own timers and were filled with a new chemical invention—napalm. The bats would be cooled in a refrigerator before the mission, putting them in a state of hibernation.

Adams planned that bombers would fly over a Japanese city just before dawn and drop the bat bombs. When each container-bomb had fallen to an altitude of 1.2 kilometres, it deployed parachutes that slowed its descent, and the sides of the device fell away, allowing the bats to fly out. As the sun rose, they would seek shelter in the nearest buildings.

The early tests used bats carrying dummy bombs. The practice runs did not go well. When the container-bomb was released and opened up, many of the bats were still in hibernation. They fell to the ground like stones.

There were worse hitches. Adams worked near Carlsbad Caverns, where he had first come up with the idea for his bat bomb. He now used the site to collect bats for his project. Six bats escaped their confinement and flew under a fuel tank in the local airbase. The bats were all "armed and dangerous" and the subsequent fire destroyed the test range.

After that incident, the air force wanted nothing more to do with bats. But Adams was endlessly optimistic and enthusiastic, and he managed to keep his project alive. The air force passed the project to the navy. The navy passed it to the marines.

Eventually, the team started to make progress. When they staged an attack on a mock-up Japanese village, built from wood like the houses in Japan, the bats did indeed find their way in or under the roofs of the houses and started many fires.

Despite a few modest successes, Project X-Ray was proving to be an expensive program—the military had spent millions developing and testing the system, and it was taking longer than they expected to produce a usable weapon. When they learned that it wouldn't be ready to use until 1945, the top brass scrapped the program. Instead of dropping a million tiny bombs on Japan, they were going to drop two big ones.

Adams regretted that the system was never used. He believed his bat bomb could have caused more devastation to Japanese cities than the atomic bombs did, but without the terrible loss of lives. Of course, the bats might have seen things differently.

EXPLODING DOGS

During World War II, some amateur inventors were very persistent in pitching their strange weapon systems.

The American animal psychologist B. F. Skinner wrote about a young man named Victor who turned up at Skinner's university lab to discuss a great new idea.

It was 1942, and once again, German U-boats posed a serious threat to Allied shipping. Victor proposed an anti-submarine torpedo that would recognize and home in on the sound of a submarine.

In those days, there was no electronic technology that could pull off this trick, but Victor didn't need electronics. He planned to train dogs to identify submarine sounds and then place them in the torpedoes as guidance systems. Victor wanted Skinner to give him a testimonial saying that the plan could work.

This idea is wrong on so many levels, but particularly in the choice of animal. If you think of creatures that navigate using sound, the first choice might be a bat, although it doesn't fly underwater. Dolphins navigate using underwater clicks—they might work for the task. Seals and sea lions also use a kind of sonar in their hunting. But dogs are a strange choice.

Still, Victor was determined to stuff dogs inside torpedoes and send them off on canine-kaze missions, and he had gone from one company to another, trying to find someone who would back the plan and get the military to buy in. Skinner had taught pigeons to steer different vehicles by pecking. When Victor learned about Skinner's work with pigeons, he was excited and included the pigeon data in his subsequent corporate pitches as an example of what animals could do.

If you're a dog lover, you'll be pleased to know that Victor's crazy ideas went nowhere, and his sea-dogs stayed on shore. However, one company was intrigued by Victor's talk about the pigeons, and they contacted B. F. Skinner to see if he could design a pigeon-based system for them. It led to a different, equally strange weapon.

PROJECT PIGEON

The military sometimes deals with bird-brained ideas, but in 1942, the US Navy developed a gliding bomb that took this further—it used three bird brains.

The bomb was a primitive drone aircraft. It had wings and a tail. It could be released from a bomber and, instead of just falling, it would glide toward its target, steered by a radar system in the nose. Such a bomb would be perfect for attacking ships—you could deploy the bomb before you came in range of anti-aircraft guns. Unfortunately, the mechanisms designed to keep the bomb on target were so large and heavy that they didn't leave much room for the actual explosive. The military needed to make the guidance system lighter.

One of the decision-makers heard about the work that psychologist B. F. Skinner had done with pigeons. They approached him about the possibility of training pigeons to steer a bomb. Skinner was surprised at the proposal, but yes, he believed it could be done.

Skinner was hired. He put pigeons in boxes with a projection of a typical bomb target on a screen in front of them and trained them to peck at the target. If the target moved, so did the pigeon, and its off-centre pecks would control other mechanisms that brought the gliding bomb back on target, like a cruise missile—or perhaps a "coos" missile.

Skinner himself described the plan as "crackpot," but to everyone's surprise, it worked quite well. The pigeons easily kept up with the movements of the target on the screen in front of them, and if something briefly blocked the target (the way a cloud might do in real life), the pigeon simply waited, then started pecking again as soon as the target reappeared.

The pigeons' abilities quickly improved because they received a reward when they did well on the test—although on a live mission, the only reward they would receive would be to die in an explosion. Nobody seemed to worry too much about that, though. Skinner commented: "The ethical question of our right to convert a lower creature into an unwitting hero is a peacetime luxury."

To prevent the possibility of a rogue pigeon steering the bomb in a creative new direction, the system used three pigeons, each with its own screen showing the forward view from the bomb, and its own set of peck-powered controls. Like the psychics in the film *Minority Report*, the birds were strapped into their chambers, and their efforts were combined to control the bomb. If two pigeons agreed, and the third had different ideas, it would be punished. The pigeons quickly learned to make obvious choices rather than idiosyncratic ones.

After much testing and refinement, the system was finally demonstrated to a group of visiting experts. It worked perfectly. And yet, somehow, the idea of pigeons controlling a powerful piece of military hardware seemed so fundamentally ridiculous that the flawless performance came across as amusing rather than impressive. The project was cancelled.

Skinner kept some of the pigeons, to see how well they remembered their training. They did well—even with no bombs to guide, they retained their piloting skills for years.

Skinner later decided that he had spent his efforts in the wrong direction: instead of using all his time changing the behaviour of the pigeons, he should have focused on changing the behaviour of the bureaucrats.

CHAPTER 20

MOBILE MASHUPS

GIVE a six-year-old a box of crayons and you're likely to get drawings of weird combination vehicles—a flying car-balloon or a spaceship-robot-steamroller.

Wherever the urge comes from to create these vehicles, it doesn't always disappear when kids grow up. The mashup vehicle is a staple for inventors in fiction and in reality. And some have even worked well.

LANDSHIPS

In the early 1900s, the public were impressed by "ironclads"—giant ships covered in armour. These big ships seemed like the ultimate weapon, huge and almost invulnerable, and several works of fiction from the day imagined similar vehicles that travelled on land. One of the best known stories is the 1903 H. G. Wells work "The Land Ironclads," but he was just adding his contribution to an existing genre. Although they were fictional vehicles, fiction sometimes turns into fact.

During World War I, soldiers found themselves pinned down in trenches, unable to advance. Military planners on both sides gave serious thought to building a real "land ironclad" or "landship"—a heavy, bullet-proof vehicle that could roll across muddy fields and break through enemy lines. The British navy was particularly interested—perhaps because landships would give them more to do. World War I had been rather light on thrilling sea action.

Early landship discussions were about vehicles that were literally the size of warships. The final product was not quite that large but still had ship-like qualities, with a crew of eight and a naval gun on each side. It even operated like a ship—the commander gave instructions to the crew, who worked the engines, gears, and brakes on each side of the vehicle, adjusting the speed of one track or the other to make the vehicle turn in different directions. The vehicle presented an extremely unpleasant environment for the crew—very hot, deafeningly loud, and full of engine fumes, but it was intimidating to the enemy.

During its development, the vehicle was kept secret. Talking about "landships" would have been a giveaway, so those in the know pretended it was a tracked vehicle for transporting water.

The landship was initially called a "water carrier," until somebody pointed out that it would be abbreviated "WC"—the British abbreviation meaning a toilet (from "water closet"). It was an undignified moniker for their new superweapon. The deceptive code name was promptly changed to "tank." That label stuck, and it has been used for tracked, armoured fighting vehicles ever since.

After the success of the tank, military minds seemed to become more open to other unusual vehicles, and over the years, a legion of inventors offered their novel ideas.

FLYING SUBS

There's an old story about a chain-smoking monk who went to his superior and asked if he could smoke while praying. The answer was "Absolutely not!" A fellow monk said afterwards that he'd asked the wrong question—he should have asked if he could pray while smoking.

Combination vehicles sometimes work the same way. A submarine that flies sounds a lot more appealing than a plane that sinks, and the "flying submarine" has certainly been a popular choice in science fiction adventure. An attractive yellow flying sub served as a shuttle on the 1960s TV series *Voyage to the Bottom of the Sea.* A plane-like flying sub was a UFO-killer in the series *UFO,* and yet another design was the submersible lair for the villain in the film *The Incredibles.*

While flying subs have mostly been confined to fiction, some were seriously considered for military use. One version that went further than most was a Russian design from a young engineer named Boris Ushakov. He knew that a submarine usually had to lie in wait for ships, and the sub's low profile means it can easily miss spotting a potential target. But what if a submarine could fly over the ocean scanning for ships? When it spotted a potential target, it would dive into the ocean ahead of its prey and wait until the ship came within range of its torpedoes.

Officials reviewed the idea but eventually nixed it. They thought—probably correctly—that a submarine plane would be too heavy to fly well. And if you're already flying in a plane, why not just attack as a plane? A single-purpose aircraft is easier to build and is bound to be lighter than a flying submarine. It can also go faster and carry a greater weight in torpedoes.

But there is no point explaining this to a rabid inventor—once they get an idea in their head, it can be hard to shake them out of it.

An American inventor named Donald Reid tried to get the US military interested in his flying submarine ideas in the early 1960s. Nobody would take him seriously, so he built one himself as a private venture and successfully flew it, then demonstrated its submarine abilities by taking it a few feet under the water.

Reid's design was primitive—the ramshackle, homemade plane rested on two floats, which could be flooded to sink it. The pilot had to wear scuba gear for a dive, and before submerging, he had to remove the propeller and put a rubber cover over the engine to keep it dry.

Of course, the military had no interest in Reid's crazy submarine plane.

That was partly because they were already working on a crazy submarine plane of their own.

The military had a contract with the aircraft company Convair to develop its own flying submarine—or, as Convair called it, "submersible seaplane." Convair's vehicle was rather more advanced than Reid's. Their engineers planned a jet aircraft designed to hunt for Soviet submarines. When it spotted a sub from the air, it would dive into the water to attack. The project was a huge technical challenge. Although the Convair plane was to be a jet aircraft, the requirements for underwater travel made it heavier and slower than most propeller planes.

Convair's efforts had a healthy budget, but that didn't help it succeed. The project was cancelled in the mid-sixties.

Proposals for flying submarines pop up every few years. So far, none has produced a useful vehicle. We're betting this is one idea that is more likely to sink than take off.

FLYING TANKS

Perhaps a flying sub was a winged washout, but how about a flying tank? In the 1930s, an American inventor named Walter Christie became interested in this concept.

It sounds a strange idea, but if it could be done, a flying tank would certainly fill a need. Regular tanks are difficult vehicles to transport. They tend to be slower than wheeled vehicles, so they're hard to send quickly to a battle zone. Any army would love a tank that its crew could fly straight into battle.

Christie was no armchair inventor—he was very experienced at tank design and had worked on several innovative ideas for the military—but his flying tank idea was the most ambitious. He wanted to give the tank its own large set of biplane wings. A tank already has a powerful engine, and with the right gearing, it could be used in the air to drive a propeller. When the tank-plane landed, it would shed its wings and rumble into battle.

Flying wasn't the only trick Christie had in mind for his tank. It would also be unbelievably fast on the ground. He envisioned a tank that could go so fast on its treads that, with wings reattached, it could take off again on the roughest ground—a regular racing tank.

Christie gave press interviews where he sang the praises of his wonderful invention, which (like many imagined weapons) was supposed to be so powerful that it would end wars before they could begin. "Knowledge of its existence and possession will be a greater guarantee of peace than all the treaties that human ingenuity can concoct. A flock of flying tanks set loose upon an enemy and any war is brought to an abrupt finish."

A 1932 issue of *Popular Mechanics* described the vehicles in action with breathless enthusiasm. "Scores of big airplanes roar majestically over enemy territory without firing a shot or dropping a bomb; they swoop suddenly to earth, then zoom into the air again, leaving on the ground the queer-shaped cockpits they had been carrying. The grounded cockpits, veritable small fortresses, move forward, seventy-five, then eighty-five miles an hour, spitting death and destruction from machine guns. One hundred flying tanks, manned by just 200 men, would rout an army of several thousand."

It sounds fantastic. Where do I sign?

In reality, the tank didn't quite measure up to the magazine's hyperbole. Only one prototype tank was built—and that was only the tank section, without the wings. The tank could certainly go fast, but that was mostly because it had been made ultralight. The prototype's armour was fake, "just to give an impression of how it would look." The military saw right through the deception—they knew that the real tank would never achieve the same race-car speeds. The project was abandoned, without ever getting to hurtle across a battlefield, spitting death and destruction and ending war forever.

The big problem with the tank-aircraft combo is that, to fly well, an aircraft needs to be light for its size, whereas what makes a tank useful is its armour, which is invariably heavy.

Still, Christie's well-publicized ideas made an impact. A Russian inventor, Oleg Antonov, took over where Christie had left off, and thought about trying to attach a tank to a set of biplane wings. His idea was simpler than Christie's—he didn't need his tank to fly on its own power. A gliding tank would be just fine. It could be towed to the battlefield by a regular aircraft.

Rather than building a new tank from scratch, Antonov started with a Soviet T-60 "light tank." It might be called light, but it was still too heavy to fly. Engineers did everything they could to reduce the load. They took away the ammunition, removed the weapons, and pulled off the headlights. They even emptied the fuel tank, to save some weight there. After all that, with a giant set of wings attached, it could just about get airborne, but it was an ungainly aircraft. Not only was it heavy, but the tank's slab-like surfaces were a serious aerodynamic challenge.

They took it up for a test flight. The tank-glider was hitched up and towed into the air by a huge heavy bomber, but the bomber's pilot soon ran into problems: the glider's weight and drag were so extreme that it risked bringing down both planes. The bomber crew panicked and cut the tow cable. They flew free, leaving the glider and its unlucky pilot to deal with his airborne anvil.

The test pilot, who was something of a legend among Russian glider pilots, actually managed to land the clumsy aircraft. He then drove the tank back to its base.

It had been a close call, and it was a lot of effort to transport a single lightly armoured tank that was only the size of a small car. The project fizzled out after that.

Other countries also experimented with flying or gliding tanks. Their results were similar.

Again, the appeal of the flying tank is all in the name. Thinking of it as a heavy, armour-plated plane gives a better idea of the challenges and why it just plane tanked.

EVERYTHING ON TRACK

What other crazy things can you do with a tank? A military engineer named Percy Hobart was determined to find out. He invented a wide range of strange armoured vehicles, mostly based on Sherman and Churchill tanks. Most of his inventions were designed for the D-Day landings.

A tank doesn't seem like it would make a good boat, but the Sherman DD was a swimming tank. A fabric "bathtub" could be raised around the upper half of the tank, making it buoyant enough to travel through water. A set of propellers at the back pushed it through the waves. The DD stood for "duplex drive," although the soldiers quickly gave the initials a different explanation, claiming it stood for "Donald Duck."

The "Crab" was a mine-clearing tank. It carried a spinning flail ahead of it, which whirled steel chains against the ground. The chains could harmlessly detonate buried mines before people or vehicles passed over them.

The "Crocodile" was a tank where the front machine gun had been replaced by a powerful flamethrower. The tank pulled a trailer carrying 1,800 litres of fuel and could hurl fire for over 100 metres. It has been described as a formidable psychological weapon—which makes sense: a person's brain doesn't work too well when it's on fire.

Hobart introduced many other strange tanks—one could turn itself into an instant bridge, another carried a gigantic spool and could lay out a temporary road made of canvas, while yet another was designed to dazzle an enemy with an extremely bright light.

This collection of odd but effective vehicles was affectionately known as "Hobart's Funnies."

SUBMARINE TANKS

Submarines and tanks are both strongly built vehicles, so a combination submarine tank seems like it might be another promising option. Such vehicles have appeared in science fiction—for example, in John Wyndham's *The Kraken Wakes*—but for some reason, this combination seems to have been less appealing to the crackpot inventor.

A 1924 magazine, *The New Science and Invention*, had a cover depicting the "submarine-land dreadnaught"—a gigantic red ship, crushing tall buildings under its massive treads as its multitude of cannons shoot their fiery wrath in all directions. If you have the GDP of a superpower at your disposal, it would be an impressive way to spend it all.

On a more modest level, in 1944, inventor Chinneth Hill patented a curious design for a submarine tank.

His was a teardrop-shaped vehicle with a small periscope, a propeller at the back, and treads at the bottom. The nose of the tank houses what appears to be a weapon, but it's not clear if it's a gun or a torpedo tube. It may be both.

The following year, another American inventor, Robert Sutphen, patented a submarine tank. His design was bigger and more streamlined than Chinneth Hill's and included two large turrets. Its tank treads were hidden inside bulges on the side of the vehicle and could be lowered for use on land.

Both these submarine tanks are classified as "ornamental design," so it's possible the inventors were thinking about toys rather than serious weapons of war. Then again, perhaps they were using the "ornamental design" category to stake a claim on what the inventor thought was a good idea of a combination submarine and tank, without having to get into the details of designing one.

But there have been some more serious inventions in this area. Submarines can travel secretly and pop up at unexpected locations—potentially a good way of launching an invasion.

During the 1940s, the Soviet Union tried to develop an enormous submarine that could emerge from the sea and crawl up onto a beach as a landing craft. While not a tank itself, it would release tanks, artillery, and soldiers onto the beach. The United States considered similar designs.

The plans for these amphibious submarines were abandoned when both sides realized that, if you wanted to use your submarine to unleash hell on an enemy, it was easier to do with nuclear missiles.

PYKRETE

For a really weird military mashup, how about a combination aircraft carrier and iceberg?

In World War II, inventor Geoffrey Pyke was working on the problems faced by ships sailing through icy Arctic waters. He turned the problem around. What if the ice was a solution instead of a problem? What if you used the ice as the ship?

Pyke's big idea was to build a cheap aircraft carrier with a revolutionary design. Instead of being welded together from steel plates, the ship would be constructed from seawater frozen into a ship-shaped block of ice. Because ice floats, the ship would be virtually unsinkable. You could attach engines to the ice to move it. Refrigeration units would keep the vessel frozen in warmer water, and its cheap, buoyant materials meant it could be built much larger than any steel ship. Such a ship would be hard for an enemy to destroy—even if it was bombed and cratered, it was a simple matter to patch the surface by pouring in more cold water and waiting for it to freeze.

Officials were intrigued, and they tested the idea using a large ice block on a lake in Alberta. A small motor ran a refrigeration unit to keep the block cold. It stayed frozen through the summer.

But it wasn't perfect. The ice was brittle and developed cracks. Pyke consulted with another scientist, who suggested that the strength of the ice could be improved if it was reinforced with wood or paper.

Pyke tried the modification, and he was amazed to discover that the cellulose fibres in wood not only made the ice much stronger, but they also provided insulation from heat and greatly slowed its melting. The new material came to be called "Pykrete."

Lord Mountbatten, who was in charge of technical innovations for the military, was excited by the results. According to one story, Winston Churchill was enjoying a steaming bath when Mountbatten burst in and dropped a block of the new ice-wood mixture into the prime minister's tub. The outside melted, forming an insulating surface of wood chips. The inside of the block remained solid. Churchill was amazed—or was his wide-eyed look just shock from the sudden drop in temperature?

In a later demonstration, Mountbatten showed a group of VIPs how Pykrete resisted bullets. He displayed two frozen blocks, one made of ice, the other made of Pykrete. He pulled out his pistol and shot into the ice block, which immediately shattered. Then he shot into the block of Pykrete. That block was undamaged—although the demonstration was spoiled slightly when the bullet ricocheted and injured a nearby official.

Plans were made to build a huge Pykrete aircraft carrier that could support the D-Day landings. It was known as Project Habakkuk, named after a biblical prophet who wrote about turning the tables on an invading enemy.

Unfortunately, Pyke's invention didn't turn the tables on anyone. It turned out that Pykrete had some problems. Although it was surprisingly good at staying solid in seawater, the stresses on a big ship are enormous, and unless the interior of the Pykrete was kept at Arctic temperatures, it tended to sag and bend.

And, while it was clear the vast ship would need enormous engines, the requirements for its refrigeration units would be even more formidable. Despite the superior insulation and strength offered by Pykrete, freezing an ice cube bigger than a supertanker would still need thousands of kilometres of metal tubing to carry the refrigerant through the ship. The refrigeration unit alone would require an unbelievable amount of steel—so much, in fact, that it could be used to build a conventional ship of the same size.

After a careful analysis, engineers realized the "budget" Pykrete aircraft carrier was a very expensive proposition—building a giant aircraft carrier from ice would cost more than building an entire fleet of traditional carriers. Although the carrier project went nowhere, the curious properties of Pykrete have remained fascinating to engineers and scientists, spawning many small-scale projects and experiments. It's a classic "solution in search of a problem."

FLYING AIRCRAFT CARRIER

Because aircraft carriers need a runway for planes to take off and land, they are often among the largest and heaviest of all military ships, so the idea of building an aircraft carrier that can fly seems a special category of crackpot idea. Yet there have been quite a few serious attempts to do just that. Most were not successful.

When zeppelins and other airships were popular, several countries tried to use them as huge, lighter-than-air aircraft carriers. In 1917, one early British airship carried a Sopwith Camel biplane strapped underneath the vessel. The idea was that the airship could launch a fighter or two to defend itself from attacking aircraft. It was a single-use aircraft—once the fighter was dropped and had done its fighting, the pilot was on his own in finding somewhere else to land. The idea was successfully tested but never made it into action.

In the 1930s, the United States had a more ambitious airship aircraft carrier program. They built two enormous airships, the USS Akron and the USS Macon. The airships had a hangar inside carrying a small squadron of fighter planes. The airships could release the warplanes, and they could also retrieve them again—each plane had a hook over its wings, and the incoming pilot needed to lock it onto a metal "trapeze" dangling below the dirigible—a delicate and dangerous operation. Once hooked, the plane could be hoisted up inside the airship's hangar.

Each of these sister ships was nearly as big as the Hindenburg but filled with non-flammable helium. However, that didn't make them any safer to fly in. Within four years of their launch, both airships met the same fate—they were caught in storms and destroyed when they crashed into the ocean. The crash of the USS Akron was particularly tragic—the flying aircraft carrier crashed into the Atlantic Ocean off New Jersey, killing most of its crew. While the Hindenburg disaster is the most famous airship catastrophe, the slow crash of the USS Akron killed more than twice as many people and is the worst airship disaster in history. Tip to any future airship builders—don't forget the life rafts.

The Soviet Union tried a different variation on flying aircraft carriers in the 1930s, as part of a project known as "Zveno." Up to five biplanes were strapped to the wings and fuselage of a heavy bomber. It looked ridiculous, but the combination did manage some successful attacks in World War II.

A similar idea resurfaced in the 1950s, when American engineers planned to put four Goblin jet fighters inside a modified bomber, the enormous B-36 Peacemaker. The fighters were tiny, each about the size of a large bomb, and would be launched and retrieved using a dangling "trapeze" lowered from the body of the big bomber—exactly the same method used on the earlier airships. The performance of the miniature jet planes was disappointing, so the military tried again with more conventional aircraft, this time connecting the plane to the "mothership" by its wing tips. In tests, the amount of turbulence near the wings made docking a very dangerous operation, so that idea was also dropped.

One of the wilder flying aircraft carriers in the 1970s used a Boeing 747 jumbo jet as its mothership. The big plane would carry ten small fighter planes, launching them from a bay in the front, retrieving them on a mechanical arm at the rear, and providing refuelling from a hose in the centre of the plane.

But an even larger aircraft had been proposed by Lockheed in the 1960s. Known as the CL-1201, it was a gargantuan jet aircraft, much bigger than any airplane ever built, before or since, and weighing around 6,000 tons. The wingspan was a third of a kilometre. It was to be powered by its own nuclear reactor, allowing it to stay in the air for more than a month at a time. It didn't carry miniature aircraft—quite the opposite; its aircraft were big F-4 Phantom fighter-bombers, and it would carry twenty-two of them. The idea was studied but never built.

Ludicrous cost might have been a factor.

ALLIGATORS OF CANADA

This section started with one kind of landship, and it ends with a very different kind—an old and little-known vessel designed to travel over lakes and through forests. Surprisingly, this successful form of amphibious vehicle wasn't developed for the military but for the Canadian forestry industry.

Joseph Jackson ran a logging business in Northern Ontario, but his work was hard and getting harder. By the 1870s, the most desirable and accessible trees, near big rivers, had been chopped down. As lumberjacks moved to more remote locations, transporting the logs to the lumber mill became increasingly difficult.

The Ontario wilderness is a patchwork of small lakes, rivers, and streams, with many isolated from each other. Loggers who cut trees around the shore of an isolated lake faced the problem of how to transport the logs to the mill. The only way was by water, but even if they built waterways, it took a lot of work to pull the logs together into a raft. They would have liked to use steamboats for their work, but no big boat could reach the scattered lakes in the middle of the wilderness.

Jackson approached John Ceburn West, an engineer from Simcoe, Ontario, and explained the problem. West came up with a plan for a new kind of boat. It would be a steam-powered tug, a flat-bottomed vessel equipped with a powerful winch. It could move around a lake using paddle wheels, but this wasn't how it towed logs—that was done using its anchor. By planting its big anchor in the lakebed, and winding in a steel cable with its winch, the vessel could easily tow thousands of logs slowly but steadily down a river or across a lake. The logs could then be transported to the next lake along narrow waterways or using a flume—a wooden structure that floated the logs down a slide.

But the clever part of the boat's design was how it reached those isolated lakes in the first place. If its winch cable was hooked around a sturdy tree, the boat could haul itself out of the water and across the land, sliding along on its steel-reinforced skids. A boat that could haul 60,000 logs had no trouble pulling its own weight.

West called his boat the "Alligator." Like its namesake, the Alligator tug looked slow and ungainly on land, but it was surprisingly adept at getting around. With a team of workers to cut a portage route, this tough little boat could drag itself a few kilometres in a day, scraping and sliding though swamps, and even up hills, arriving at the next lake in time to rejoin its cargo of logs and pull them to the next stage in their journey.

The Alligator tugs were a huge success, and West's company, West and Peachey, manufactured hundreds of the vessels, selling them to timber companies in Canada, the US, and even South America.

What eventually led to the extinction of the Ontario Alligator was another engine technology. Smaller, cheaper diesel engines allowed logging companies to send trucks more easily into remote areas and to set up small sawmills close to where the trees were being felled. The last steam-powered Alligator tugs were hauled up onto beaches and left there to rust and rot away. Sadly, the breakthrough technology is now mostly forgotten.

CHAPTER 21

A VERY TABLE GENIUS

EVEN the smartest people need to eat and drink. But that's never stopped them from creating. Some have even used these breaks to reimagine the way the rest of us have our meals.

Their inventions are designed to make dinner more efficient and protect us from table dangers like . . . the grapefruit?

GRAPEFRUIT SHIELDS

It's surprising how many inventors have tried to protect diners from grapefruit juice.

In 1928, Joseph Fallek invented a grapefruit shield. It looked like the canopy of a baby carriage. It was mostly made of waxed paper, attached to the rim of a half grapefruit with a metal belt, fixed to the grapefruit skin with several sharp needles.

The daredevil eater could then consume the fruit without nearby diners being splashed by its spray—although they might have been at some risk of being stabbed by the needles.

Inventor Joseph Gibson also fretted about the atomized liquids. He wrote: "It is well known that in eating such fruit as oranges and grapefruit, the manipulation of a spoon therein causes the juice to be ejected and sprayed to great disadvantage, inconvenience, and waste."

His invention, from 1919, was a glass disk that covered the top of the grapefruit like a coffee cup lid. A slit at the front allowed the careful access of a grapefruit spoon. The spoon operator could look through the glass to manipulate the spoon into the depths of the grapefruit, then extract it when they were ready to eat another morsel.

Grapefruit shields have continued through the years. A 1975 invention by inventor Jack Orenstein looks like a Star Trek prop. It was a clear hemisphere resting on three legs. Placed over the grapefruit, it provided a barrier to upward spray.

It's strange that people who can happily squeeze a lemon with their bare hands, or peel and eat a tangerine, treat grapefruits as if they are filled with acidic alien blood.

Then again, as we'll see, the grapefruit does have an unusual history...

ATOMIC GRAPEFRUITS

We don't normally associate inventions with plants, but crops can be patented, and a British inventor named Muriel Howorth went out of her way to create new fruits and vegetables. Her efforts led to the modern grapefruit.

The grapefruit wouldn't exist at all if it hadn't been for humans. It first appeared by accident. In the 1700s, owners of one plantation were growing two different citrus fruits from Asia—sweet oranges from China and pomelos from Indonesia. Oranges are tasty but not very large. Pomelos are much bigger, but when you cut them open, they are mostly white rind, with a small amount of edible fruit near the centre.

One day, a rogue bee must have visited each type of tree, because an accidental hybrid appeared—it was large, like a pomelo, but filled with juicy fruit like an orange.

The new fruit grew on trees in clusters, just like a grape. People named it the grapefruit. It was a strange choice, if you think about it, because there was already a fruit that grew in clusters like a grape, and that was the grape. But nobody objected, or if they did, nobody cared. The name grapefruit stuck for this fruit that wasn't a grape.

The demand for grapefruit steadily grew. By the early 1900s, many huge grapefruit orchards had been planted in the southern United States. That's where a second genetic accident occurred.

In 1910, one mutant tree produced grapefruits that came out pink rather than the usual yellow. The new pink fruits were sweeter than yellow grapefruits and could be sold as a luxury item. Sales took off, and soon the orchards were full of them.

Another mutation appeared a couple of decades later, this time on a grapefruit orchard in Texas. The flesh of this mutant grapefruit wasn't pink but red. This was so exciting, and so valuable to the owners, that it was it was the first grapefruit to be patented—they named it the "Ruby Red."

The original grapefruit and the valuable new strains that turned up later were all natural mutations. But mutations don't happen very often, and useful ones are even less common. Wouldn't it be great if people could speed the process up a little?

This is where inventor Muriel Howorth enters the picture. Although she didn't have a formal scientific background herself, she was excited by science, and particularly enthusiastic about peaceful uses of atomic energy. Rather than fearing the atom, she wanted more people to embrace it. Some of her efforts were very eccentric—she staged a dance titled "Isotopia," where women twirled across a stage as subatomic particles.

One of Howorth's initiatives was something she called "atomic gardening." The idea was to speed up the slow processes of evolution. She would give natural selection a kick in the pants by blasting crops with radiation. She founded the Atomic Gardening Society, which she claimed had "a thousand mutation experimenters." Her ideas dovetailed with an American "atoms for peace" movement in the 1950s.

Plants were irradiated in a "gamma garden." This was a field laid out in a circle. In the middle was a rod of radioactive cobalt-60, which gave off an invisible glow of deadly gamma rays. When the rod emerged from its protective chamber, the whole area was irradiated, and nobody could enter the "garden" unless they were wearing protective gear. Laid out around the radiation source, like the segments of a grapefruit (or a radiation warning symbol) was the garden, with a different crop in each wedge.

The gamma rays played havoc with the plant DNA. The plants nearest the centre received so much radiation that they usually died. Further out, the plants were misshapen or had plant tumours. But on the outside of the circle, plants grew more normally and sometimes produced useful mutations.

Seeds from these gardens were sent to members of the Atomic Gardening Society, who grew them and took notes to see what would emerge. It sounds strange today, but many people in the 1950s and 1960s were excited about the prospects of these modern, high-tech farming methods.

The method is not quite as dangerous as it sounds. Although the plants and seeds were exposed to radiation, there was no radioactive material inside them, and they were not radioactive themselves.

Howorth promoted the gamma garden, and partly thanks to her influence, many food growers became interested in giving this new technology a try.

Down in Texas, the grapefruit industry wanted to create even redder, even sweeter fruit without waiting decades for the next mutation. They blasted radiation at grapefruit plants. It worked like a charm. The redness comes from a chemical called lycopene, and when they blasted gamma rays at their crops, some of the mutants were loaded with it. The strains of red grapefruit that came out of these experiments are now some of the most popular grapefruits in the world.

It may not be an accident that lycopene not only makes fruit red, but also protects cells against radiation damage, so these grapefruits, created with radiation, should also be the most resistant to it.

PASS THE EGG SPOON

In Victorian times, an obsession with table manners led to the invention of dozens of new forks, knives, and spoons.

Etiquette writers were like modern "influencers"—they offered advice to a new and nervous middle class, hoping to climb the social ladder. The writers' opinions not only sold books, but could also be profitable for tableware manufacturers. Most people owned their own set of spoons, but if the books told them they needed a grapefruit spoon (a narrow spoon with a serrated cutting edge) in order to serve grapefruit to their guests, they would buy a set, just to show they were in the know.

For centuries, people had used a large-ish spoon to eat soup or soft food. But now it suddenly became necessary to have a dessert spoon for your pudding, and a different, rounder soup spoon for your soup.

All it took to make a new spoon was a label. Manufacturers soon started inventing new sub-types of soup spoon—a small bouillon spoon for thin soups, a medium spoon for cream soups, and a large one specifically for chowders.

An egg spoon was invented just for eating boiled eggs. Somebody must have got a bonus for that idea. A mustard spoon was for putting mustard onto your plate. A Stilton scoop was a flat, shovel-like spoon specifically for serving pieces of Stilton cheese. There was another spoon, long and narrow, solely for scooping the marrow from bones.

Then there were the forks. For most of history, people ate with their hands, and when forks were introduced, they were used mostly for holding meat while it was being carved, or to serve it to guests, who would then pick it up with their hands. But as time passed, the hands-on approach suddenly became a huge no-no. All food had to be moved with a fork. Diners needed a fork each, and by Victorian times, it had to be the right fork. When you sat down to dinner, you needed to identify the bread fork, the fish fork, the snail fork, the olive fork, and even the mango fork. ("Oh, one of *those*!")

Some new pieces of cutlery were complicated and specific. Asparagus tongs, as the name suggests, were just for moving asparagus from a serving dish to a diner's plate. Grape scissors were a pair of long, ornate scissors with the sole purpose of snipping grape stems.

Anyone seeing the array of table hardware for the first time would probably be confused and overwhelmed about what to do with them all—but that was the whole point. Manners drove this inventive bonanza. Each item at the table had its own rulebook, and an elegant meal also served as a secret test of how up-to-date people were on the latest complex code of polite behaviour, as well as showing off the host's superiority.

Most of this flatware was not born of necessity. Nobody ever said, "Finally! A way to lift a piece of asparagus!" Instead of offering a solution to a problem, these pieces of cutlery posed a problem, and the cultured diner was supposed to know the solution.

Today, people are less obsessed with etiquette. We routinely eat many foods with our hands again—sandwiches, burgers, wings—and there's less need to spend a large portion of the family's wealth on a complete set of silverware that is never quite complete enough.

Although most of the cutlery invented by the Victorians has disappeared from general use, a few are still going strong. We still have cake serving knives and steak knives, as well as the odd serving fork.

The most successful of all the new tableware was probably the teaspoon. It was originally introduced for the highly specific job of putting sugar into your cup of tea, and it's still used that way, but it's also become the general-purpose small spoon, used for everything from measuring ingredients to eating yogurt from the tub.

The Victorians would have been appalled.

Fortunately, social success no longer depends on being scrutinized over the correct use of arbitrary cutlery. Today, we have more freedom in how we eat—as long as we start by posting gorgeous images of the meal to feed our followers' envy.

BEER WITH A TWIST TOP—AND BOTTOM

Some people disapprove of drinking beer straight from the bottle, even in casual situations. But if that's how you feel, what do you do when there are bottles but no glasses?

In 1900, William Baum of Sullivan, Missouri, wrestled with this problem, trying to come up with a way of bottling beer so that a thirsty person with no glass could enjoy a more refined style of drinking.

On the face of it, it doesn't seem like a difficult problem. Perhaps the narrow-necked beer bottle could be replaced with a wide-mouthed jar.

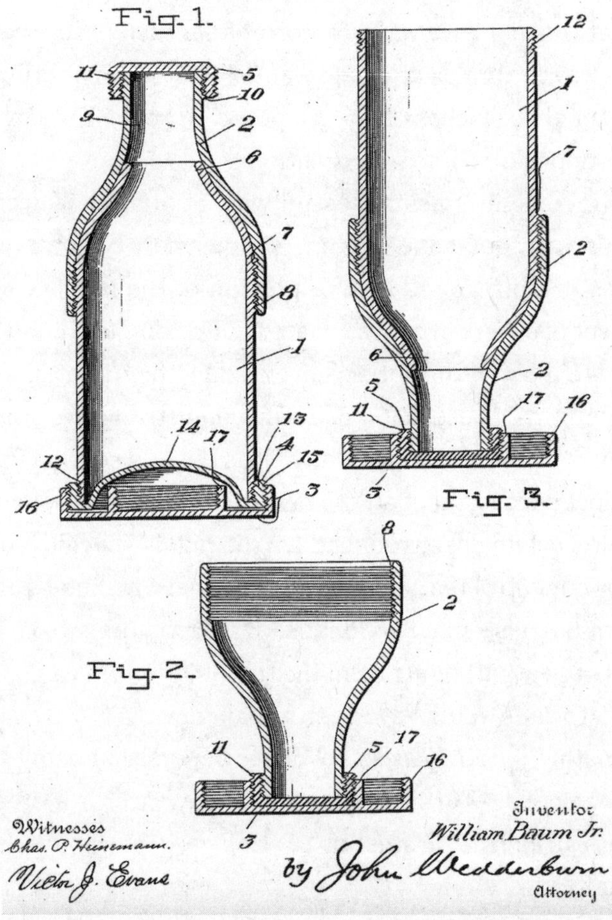

But that sort of solution was too easy for Baum. Instead, he designed a strange drinking vessel that looked like an ordinary beer bottle but could be ingeniously reassembled like a James Bond gadget.

First, the base of the bottle unscrewed. The beer wouldn't leak out because it was a false base, with a second base underneath. The neck of the bottle also unscrewed, again revealing a second, slightly shorter neck below.

The false neck and base could then be screwed together—the narrow end of the neck flipped upside down into the base to form a small goblet resembling a wine glass.

The user poured beer from the remainder of the bottle into this small drinking vessel, which would look very dainty, assuming the beer didn't dribble out from the glass-on-glass connection at its base.

It's not surprising Baum's invention didn't catch on. It's a complicated construction, and delicate, with multiple levels of glass threads screwing into more glass threads. From a manufacturer's point of view, it would have been cheaper and easier to bundle a free glass with every bottle.

THIRST FOR KNOWLEDGE

In 1885, Herbert Jenner of Washington, DC, came up with a method for hiding alcohol rather than serving it. His patent for "improvements in liquor flasks" is a fake book containing a small bottle of booze.

In the patent's illustrations, the spine of the book is given a deliberately dull title: "Legal Decisions, Vol. II." Apparently, it's less about the letter of the law and more about the spirit.

So far, this invention doesn't seem very remarkable—secret drinkers have always hidden their bottles in drawers and boxes, and there's nothing new about hiding valuables in a hollow book.

But Jenner's invention improves efficiency with a fake book designed specifically for bottles. The top of the book has a small trap door situated over the neck of the bottle. If you want a drink, just press your finger into a hole at the bottom, lifting the bottle so its neck sticks out through the trap door. Then unscrew the lid and drink. No time is wasted opening the book and removing the bottle.

And if someone enters the room in mid-slurp, they won't see you drinking from a bottle of liquor but will only see a person innocently sucking on a raised book of legal decisions. Okay, maybe that part doesn't work so well.

BOTTLE BRUSH

If hiding your booze inside a fake book seemed too risky, there were other options. In 1893, Thomas Helm of Danville, Virginia, patented a combination clothes brush and bottle.

The brush looks like a large, old-fashioned clothes brush, and functions like one, too, but the thick handle is hollow and made of metal. One end pulls off to reveal the neck of a small flask built into the handle. The removed end piece can serve as a drinking vessel.

The brush could be placed on a table, hiding the alcohol in plain sight—although it does carry the risk that anyone using the brush to clean a coat might notice the strange sloshing sounds coming from the interior of the brush.

But barring an accidental discovery, the owner can sneak a drink from their clothes brush anytime in complete safety—and enjoy the feeling that they've hit rock bottom.

BURP-BUSTING BABY BOTTLE

From bottles for alcohol, we turn to bottles for babies.

In 1953, in Drummondville, Quebec, a sixteen-year-old named Jean St-Germain was babysitting one of his brother's children and became curious about the need to burp the baby after it had drunk milk from a baby bottle.

It gave him an idea for a baby bottle that would stop the baby swallowing so much air. Instead of putting the milk in a bottle, it was stored in a plastic bag that would deflate as the baby drank. Now, when the baby sucked, it got only milk, not milk and air. And because the plastic bag was disposable, there was less bottle washing between feeds.

He sold his invention to an Ottawa businessman for a flat fee of $1,000.

The businessman made a smart purchase—the idea was resold to the Playtex company, who turned it into a commercial product in the 1960s. Their milk bottle with disposable bags became a bestseller and is still available today.

St-Germain had made $1,000 from an idea that went on to make millions, but according to his family members, he didn't regret the sale. He appreciated the money he had been paid—which was a healthy sum in the 1950s. It stimulated a taste for invention that continued all his life.

As we'll see later, he went on to become one of Canada's most creative and eccentric inventors, turning out inventions in a wide range of fields, from flying machines to robotic restaurants.

CHAPTER 22

AROUND THE HOUSE

THERE are inventions that have made home life easier, safer, and more comfortable—examples include the air conditioner, the smoke alarm, and the microwave oven.

But for every useful addition to the home, there have been scores of failed wannabes—domestic accessories that either didn't work or tried to solve a simple problem in a convoluted way. Here are a few modest examples.

AIR CHAIR SCARE

Sure, rocking chairs can help you relax, but using a rocking chair uses muscular energy and that generates heat. A person sitting on a porch on a hot day might find the heat of rocking while sitting in the sun detracts from the relaxing effects of the chair.

That's why, in 1869, Charles Singer invented the air-cooled, reclining rocking chair.

The chair was fully adjustable. Its armrests had a series of teeth below them, which could lock the back of the chair into different reclined positions—although doing so would likely throw the rocking chair out of balance.

But its most unusual feature was its built-in cooling system. The rocking mechanism was linked to a pair of bellows underneath the chair.

As you rocked, the movement of the chair pumped the bellows, creating a flow of air. A metal pipe ran from the tip of the bellows, up behind the back of the chair, and curved around to blow a cooling breeze into the user's face.

The head of the air pipe was styled like a snake's head—because nothing relaxes a person like staring up into the open mouth of a hissing snake.

A CHAIR THAT SUCKS

There were other inventors who thought about exploiting the untapped energy potential of the rocking chair.

In 1909, the Behringer brothers, Emil and Herman, wanted to break down the barrier between work and play, harnessing the power of relaxation to do useful work.

Their design had some similarities with the Charles Singer's chair, also using a set of bellows placed under the rocking mechanism, but instead of the sitter using the movement of the chair to blow air into his

own selfish face, their rocker was linked to a vacuum cleaner tube. While one person relaxed and rocked, another could use the suction created to do the housework.

It sounds ideal for couples where one needs to bustle around and work while the other is trying to relax. An illustration accompanying the patent shows an annoyed-looking woman vacuuming a living room while a man sits in the chair reading a newspaper—presumably while rocking furiously to power his wife's cleaning efforts.

The chair was not a commercial success, which is a shame, because the concept could have been taken so much further. A bank of pneumatic rocking chairs, perhaps powered by the elderly, might have run tire pumps, factory machinery, or mechanics' tools. The workers and business owners gain free energy, while the old people can feel like an important part of society, rather than frittering away their final years gently rocking on a sun-dappled porch.

WASHSTAND WRITING DESK

In the late 1800s, several inventors came up with combination writing desk and washstand. This piece of furniture looked like a writing desk, but contained a washstand hidden inside.

One 1882 model, from Nathan O. Bond, of Virginia, had a hinged lid. When you lifted it, a wash bowl would rise up from below. You could fill the bowl from a small tap—a water tank was hidden in the rear of the writing desk—and when you'd finished washing and closed the lid, the bowl tipped downwards, emptying the used water into a tub below.

It's very clever, although the curved underside of the desk looks disturbingly like a modern toilet.

Aaron G. Thayer of Kensington, Kansas, came up with a combination washstand and writing desk that mounted to a wall. The small desk with a hinged top looks like something you'd find in an old school. The top flipped up to reveal a mirror on the underside, along with places for storing pens, ink, soap, and comb. Lifting another flap revealed a small basin for washing and supplies for shaving.

Elisha E. Everitt of Philadelphia offered a more comprehensive model. It was a floor-standing cabinet, fully functional as a writing desk, with extra drawers and doors for its washstand duties. One compartment held a water jug, while another had a container for the dirty water. More compartments held a soap dish, towels, brush, and comb. Again, the desk lid flips up to reveal a mirror on the underside. Some drawers held writing materials, while others stored shaving equipment. A door in front of the writer's knees could be opened to reveal a chamber pot.

Everitt wrote: "This piece of furniture will not only serve the double purposes of a desk and washstand, but will be adapted to stand in places, such as offices, where an ordinary washstand would not be tolerated."

We suspect that visitors who wouldn't tolerate a mirror and basin might be even more alarmed about office workers relieving themselves at their writing desks.

BATHTUB WASHOUTS

Ever since Archimedes shouted "Eureka!" bathtubs have been a popular spot for inventors to do their thinking. And for some, the thinking was all about bathtubs.

A few have invented portable bathtubs. In Chapter 4, on combination inventions, we looked at Ethelbert Watts and his combination bathtub and suitcase from 1876.

In the same year, inventor Arnold Seligsberg also patented a portable bathtub. His version was designed like a carpet bag, with soft sides. It was made from rubber-coated fabric so it could unfold. Once a set of wooden supports were wedged into the sides and top, it took on the shape of a traditional steel tub. A plug on the inside, with an attached chain, allowed the bath to be drained through a hose in the front.

The inventor claimed that "it may be conveniently carried by travellers with the same facility as other light portable baggage or packages, and it affords ready means for bathing at any time in any place where water can be obtained in any manner."

It looks like you could bathe in it just like any other bathtub—although there are some obvious risks. The plug is a likely weak point—if you're in the habit of pulling on the chain with your toes, it would send the water pouring all over the carpet. Nudging one of the wooden stays out of place was likely to collapse the side of the tub, throwing the bather out with the bathwater.

The patent for a 1904 portable bath from Adolf Herz of Vienna seems like it would have done a better job of keeping the liquid in one place. The user filled a miniature paddling pool full of water, then pulled its sides up over their body, fastening it like a collar around the neck. Once it was set up, the "bath" formed a kind of tent with the bather's head sticking out of the top. Encased in what was basically a waterproof bag, they could then wash themselves, standing, sitting, or even lying down. A series of interior pockets held soap, washcloth, and other bathing accoutrements. It was ideal for the person who wanted to bathe in front of guests while hiding their nudity from their visitors' eyes and their own.

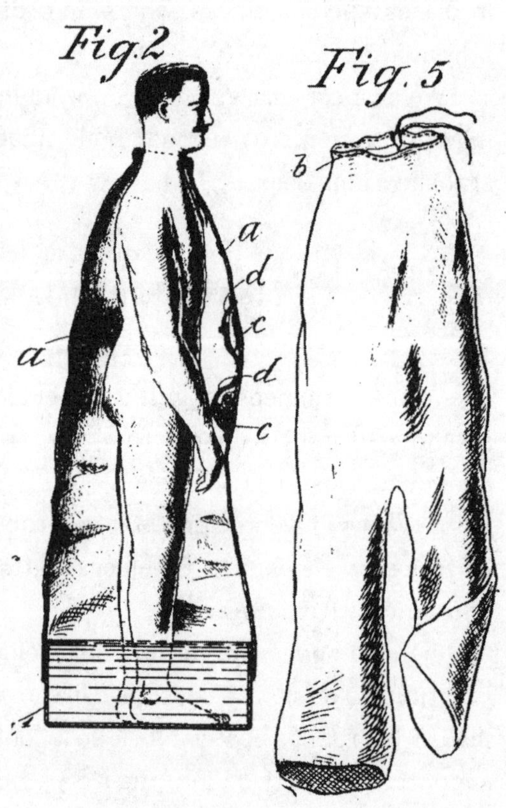

Of course, moving around the room with your feet in a paddling pool might have posed a challenge, probably requiring a series of ungainly hops. But Herz had considered this problem and offered an alternate version with two watertight trouser legs. With the two-legged version, the bather could easily walk around, sloshing two legfuls of bathwater wherever they went. It's like a strange inversion of Traugott Beck's life-preserver suit.

In 1972, an American woman named Frances Allen invented a portable bathtub. The pictures accompanying her patent show a device resembling a gigantic hot water bottle with a woman's head at the top. The container was zipped up and fastened around the neck. Water would enter through a hose at the top corner and drain through a hose in the bottom corner.

Allen's invention was designed to give a bed bath to patients in hospitals and nursing homes, as an alternative to the usual sponge bath. It would probably keep the water contained, but it increased the work for the staff—to scrub the patient, the nurse was required to push on the outside of the container, blindly rubbing a towel-like inner layer against the bather's skin.

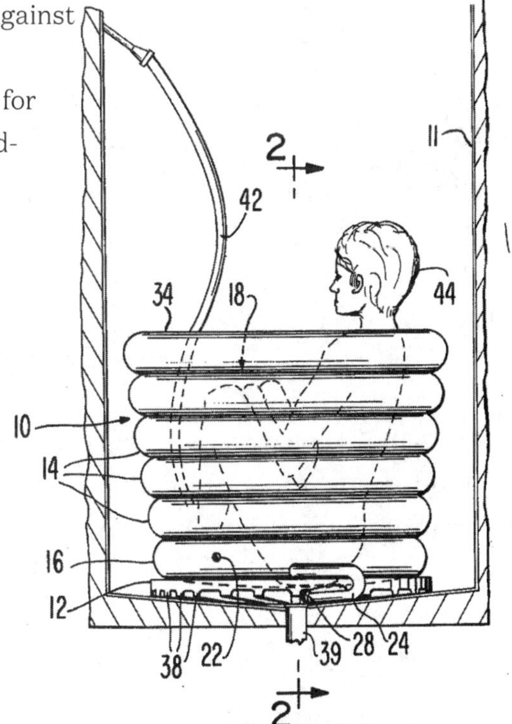

Most of the oldest patents for portable bathtubs were products aimed at travellers who would otherwise have no access to bathing facilities. In more recent decades, hotel guests might face a different problem—a hotel room that has a shower but no bathtub. Guests who have ended up in such a room but are determined to bathe in a tub could use an invention from 1977 by

Susan Younker—it was an inflatable bath, designed to fit a shower stall and resembling a stack of inner tubes. Instead of showering in a cramped stall, the user could enjoy a bath in a cramped tub.

SPIDER SAVER

Not every bathtub invention is for the benefit of humans. Kindly Edward Doughney, of Bedfordshire in the UK, was concerned about the welfare of the common house spider. All too often he had observed these creatures falling into bathtubs and being unable to escape—their hairy little legs don't have enough of a grip on the tub's ceramic surface to climb up.

In 1994, he patented his solution—a rubber ladder, complete with tiny rungs, that could be stuck to the inside of the tub. The top of the ladder fastened with a suction cup, and the rest of the ladder hung down into the tub.

If all went according to Doughney's plans, a spider that had taken a nighttime tumble into the tub would run around the bath and would eventually find the ladder (the bottom of the ladder is wider than the top to make the discovery easier). It should then be a simple matter for the arachnid to leave the bath—and take a little trip to the homeowner's bed.

THE MAD HATTER'S BED

It's said that the pictures of the Mad Hatter from *Alice's Adventures in Wonderland* were modelled after Theophilus Carter, an eccentric furniture dealer in the city of Oxford, who used to stand in his doorway wearing a strange combination of clothes—a workman's apron and a gentleman's top hat, pushed far back on his head.

The story around Oxford was that Theophilus Carter was an inventor, and one of his inventions, displayed at the Great Exhibition in 1851, was a mechanical bed with a built-in alarm clock. When it was time to get up, the bed would tip you onto the floor. Better be ready on that snooze button!

The catalogues of that exhibition still exist, and they don't mention any Theophilus Carter, but there really was at least one tipping "alarm bed" on display at that exhibition. It was the invention of Theodore Jones, who claimed that the "movement of the hand of a common watch will turn anyone out of bed at any given hour when attached to this bedstead." An image of the invention shows a bed like a giant cradle, with a shocked man being tipped out of it at eleven o'clock—although it's not clear if the picture is an accurate drawing or a cartoonist's satire.

But some people have a difficult time waking up in the morning, and the alarm bed solution has been proposed by many inventors.

In 1885, Adolph J. Nordmann of San Francisco designed an alarm bed with a clock built into the headboard. When it was time to wake up, the head end of the bed dropped down, awakening the sleeper. A whole new way to wake up with a kink in your neck.

Another, from 1892, is more like the Theodore Jones bed. The design, by George Q. Seaman of Brooklyn, connects the alarm to a mechanism beneath the bed. When the alarm sounded, the machinery pulled the support from one side of the bed, and the sleeper was rudely dumped on the floor. A person sharing a bed with someone heavy would need to make sure they were on the far side, or they might end up crushed under their partner.

Not all alarm beds deposited sleepers onto the floor. One of the odder beds in this odd category was Samuel Applegate's 1882 patent. A frame was mounted over the pillow area of the bed, hanging ominously over the occupant. It held sixty corks on strings, in a six-by-ten arrangement. When it was time to get up, a mechanism attached to a nearby clock would release the corks onto the sleeper's head.

The invention raises several questions. If a person is so fast asleep that they can't be woken by a ringing alarm bell, why would they be expected to wake when sixty corks fall on them? And why corks? And why sixty?

William Garris of Pennsylvania thought it was better to shake sleepers awake rather than knock their beds from under them. His 1905 patent is for a sturdy wooden box mounted beside the bed. When the alarm goes off, a mechanism turns a crank. The handle is attached by a strap to the sleeper's arm or ankle and tugs on the limb until the clockwork runs down.

Going to sleep with your leg strapped to a machine probably wouldn't have been a comfortable way to nod off, but the device seems like it would wake a sound sleeper. And if they tried to get up in the night and forgot about the strap on their ankle, the fall and smashing of furniture would wake everyone else.

Many alarm beds operated by shaking or dropping the dozing occupant. But Joseph Leal of San Francisco reasoned that if you wanted a person to get up, you should make them do just that. With his bed design, the alarm triggered a system of springs and weights, raising the head of the bed ninety degrees and shifting the sleeper into a sitting position. Good if you're sleeping on your back, but maybe less so if you sleep on your side or your stomach.

There are many more patents in the alarm beds category, but we're guessing that more alarm beds were patented than sold. Even if the inventions had no connection to the Wonderland character, most of them are mad as a hatter.

WALLPAPER PLAY

Sometimes an invention for the home starts well, then things go wrong as times and fashions change around it. This kind of social shift produced one very successful product.

In 1933, a soap manufacturer named Kutol was in trouble. The world was in the middle of the Great Depression, and their soap products weren't selling the way they used to. The CEO, Cleo McVicker, suggested that the company should launch a wallpaper cleaner.

Back then, most homes had wallpaper on the walls, but the paper quickly became stained with soot from coal fires, oil lamps, and cigarette smoke. You couldn't wash wallpaper—the water would soak through and make the wallpaper peel off. Cleaning it required something that would lift out the dirt but not be too wet. Professional wallpaper cleaners (yes, that was a job back then) used a product made from a mix of wheat flour and other chemicals—it was a squishy, dough-like substance you could roll against the delicate wallpaper to remove the stains.

Cleo McVicker gave his brother Noah the assignment of developing their new product. Noah mixed up his own formula of cleaning goo—a blend of flour, water, detergent, and some oil products to cut the soot. It was a soft, off-white paste and was marketed under the name Kutol Wallpaper Cleaner.

The product sold well and saved the company.

But, as the years went by, a couple of new problems appeared. The first was that homes were getting cleaner. As people switched to gas and electricity, there was less soot in the air, and that meant fewer stains on the wallpaper. Also, the wallpaper itself had changed—newer types had a vinyl coating that could be easily wiped clean with soap and water. There was no need for a special cleaning paste. By the mid-1950s, nobody was asking for wallpaper cleaner anymore.

A teacher named Kay Zufall was an in-law in the McVicker family. She ran a nursery school and thought up craft activities for the little kids. She wanted to have them make ornaments, but the children had trouble manipulating clay—it was too stiff for small fingers. She had read that wallpaper cleaner was a good craft material, so she used the Kutol brand. The kids loved it. Knowing that the company was struggling with sales, she came up with a great idea—why not forget about cleaning wallpaper and sell the product solely as a craft material?

Noah McVicker loved her idea, and they mixed up a modified formula, taking out the detergent and adding rainbow colours and a pleasant scent. At Kay Zufall's suggestion, they called the new product Play-Doh.

The company took the reinvented product and arranged a deal with children's performer Bob Keeshan, who played Captain Kangaroo on TV. He was no stranger to reinvention himself—he had played Clarabell the Clown on *Howdy Doody*, until he'd been fired by Buffalo Bob and had to invent a new persona of his own. Keeshan loved the product and highlighted it on his show.

With Captain Kangaroo behind it, Play-Doh quickly took off. Since its launch in the 1950s, more than three billion cans have been sold—a lot more than was ever sold as wallpaper cleaner.

A number of inventive processes were involved in creating the product, but perhaps the most important was the insight of Kay Zufall that a secondary use of the product could be the star feature. Sometimes invention is a lot like Play-Doh itself—it's about taking something that's already in front of you and kneading it into a new form.

Once she shared the idea, the new use seemed obvious. "Play? Doh!"

CHAPTER 23

HELICOPTER PATENTS

LEONARDO da Vinci was fascinated by flying machines. He drew pictures of an aerial screw, with a helix-like wing, powered by muscular men. It might have worked as a toy, but it wouldn't have worked to lift people.

Despite the impracticality of this design, it provided a name for a new vehicle: the word "helico" comes from the Greek for a helix or spiral. The word "pter" (as in pterodactyl) meant wing. Put them together and you get helico-pter—although now we've rearranged the last two syllables, so new words for hovering vehicles are not "pters" but "copters."

Long after Leonardo, the idea of the helicopter continued to attract quirky artist-inventors.

BALSA WOOD HELICOPTERS

When you're trying to invent a new aircraft, a mistake can get you killed, as we've seen in many ways. But one legendary aircraft inventor achieved success by crashing his vehicles again and again.

Arthur Young had always dreamed of flying. He grew up in the 1920s, a time when planes were becoming faster with each passing month. But Young wasn't interested in speeding from place to place. He wanted to build a helicopter.

Many people had tried to build practical helicopters—perhaps even more than the number who had attempted winged aircraft. Even Thomas Edison had tried. They had all failed. Generally, helicopters didn't work. It was crackpot territory back then.

None of that discouraged Young. His approach was systematic and practical. He spent years on his parents' Pennsylvania farm, building model helicopters and trying to get them to fly. His models were modest—most were made from balsa wood. Many were powered by rubber bands. He built countless designs, observing how they flew and how they crashed. It was a great way to learn about a new kind of flying machine.

Helicopters pose some unique problems. Unlike a plane, a helicopter's engine needs to lift the vehicle's entire weight. That takes a very powerful engine attached to huge rotor blades. All those heavy spinning parts make the vehicle shake and vibrate—the vibration is much worse than you'd get from the small propellers on a plane.

By the 1930s, some companies were building helicopters that could fly, but they were treacherous aircraft. Many prototype helicopters literally shook themselves apart on the tarmac. The dangerous vibration also caused rapid metal fatigue in flight. Components could snap unexpectedly. Young quickly learned how important it was to reduce vibration.

Another big problem with helicopters was stability—Young found that it wasn't too hard to get a model helicopter into the air, but once it was there, a little jolt would often send the helicopter swinging back and forth like a pendulum, taking bigger and bigger swings until the whole thing flipped over and hit the ground. The same thing happened with piloted helicopters—if the pilot's attention lapsed for a moment, the craft could swing out of control. Piloting a real helicopter from that period was a terrifying job.

After many experiments with his models, Young came up with an interesting solution to the stability problem—he put a pair of weights at right angles to the helicopter's blades. The spinning weights, later called a stabilizer bar, worked like a gyroscope—as the rotor spun, the bar would try to stay level. If the helicopter started to tip, a mechanism attached to the stabilizer bar adjusted the angle of the blades, making it level again. It worked perfectly, and Young could easily make his helicopter models hover in one place.

In 1941, just before America entered the war, Young shopped his helicopter models around, seeing if any aircraft companies were interested. He met with the Bell Aircraft Company, founded by aircraft designer Larry Bell.

When the people at Bell Aircraft saw his models, they were astonished. Young's helicopter could hover like a dragonfly. It was completely stable. They could see at once that they had a helicopter genius in their midst. They hired him, and he became one of their most important people.

Countless others had worked on their own grand helicopter ideas and failed. Young had worked mostly alone, but because he kept his experiments on the modest scale of balsa models, he could make changes and try daring ideas. His approach allowed him to fly (and crash!) far more designs of helicopter than engineers working in big companies, giving him deep expertise.

But, as we'll see, Arthur Young also made one big mistake.

A WHIRLING MIND

Sometimes it takes an unusual mind to solve a difficult problem. But that same unusual mind can also be pulled down other rabbit holes.

Arthur Young had never been a typical engineer. He grew up on the family farm, but his family were artists, not farmers. Young himself was deeply interested in philosophy, consciousness, and the universe. That's why he loved the idea of floating motionless in the sky, with nothing between himself and the ground. It was this dream that made him want to create a helicopter, not, say, a fighter jet.

When Arthur Young joined Bell Aircraft, he signed over his helicopter inventions to the company. He was then assigned the task of designing a new, small helicopter. Larry Bell could see that Young needed to work in his own way, so he was allowed to choose his own team, and he had a fair amount of independence.

In the meantime, the company's top engineers took Young's breakthrough ideas, especially the stabilizer bar, and excitedly went to work on what they considered their flagship project—a helicopter that could replace the family car.

At the end of World War II, people believed that once manufacturers had figured out how to build reliable helicopters, regular cars and planes would be history. People would travel everywhere by helicopter, launched from their own driveway. The Bell company's best minds were busy designing the "helicopter of the future."

Meanwhile, Young and a small team were kept busy tinkering on their small helicopter. It was an ugly mass of tubing. It landed on metal bars—Young didn't see the need for wheels on an aircraft that lands vertically. The front of the vehicle was completely open to the air—it was just what Young had dreamed of—you could sit at the joystick and peer down at the ground, without a window frame getting in the way.

Although his bosses generally let Young pursue his weird ideas, the idea of an open-air helicopter where the pilot could just lean forward and fall out was a little too weird. He was ordered to put a canopy around the pilot.

Young was disappointed, but he started thinking about how he could add a windscreen while keeping that magical sense of floating in the sky. What if the windscreen was a big bubble? That would be cool, but how do you make a giant clear bubble?

He had a wild idea. What if you blow it like a soap bubble? He took a giant sheet of Plexiglas, heated it up, then blew air into it. It inflated into a beautiful plastic bubble. When his engineers tested it, they found that it was thin but strong, and it was crystal clear. It was a brilliant solution.

They stuck the bubble on the front of their aircraft. The tail was an open frame of metal tubes.

In 1945, the company unveiled its new machines. The star of the show was the "flying car," now called the Bell Model 42. It was a truly beautiful helicopter—its body was sleek and rounded with a teardrop shape. It had five seats just like the family sedan—room for a pilot and four passengers. It landed on three big wheels like a plane. It had wide, car-style doors, a plush interior, and plenty of chrome. It had been pitched as the helicopter of the future, and that's exactly what it looked like.

But for some reason, visitors were more intrigued by the ugly duckling parked nearby. Young's helicopter, with its framework body, landing skids, and strange bubble canopy was like nothing anyone had ever seen before.

When pilots tested the vehicles, they were even more impressed by the ugly duckling—it was a funny aircraft to look at but a delight to fly, responding perfectly to the pilot's movements. And, of course, the view looking up or down through that huge bubble canopy was simply incredible.

As for the showpiece Model 42, it might have looked like a racehorse, but it flew like a cow. The controls were slow and heavy—sometimes it took two people to fly it. Unlike Young's team, its engineers hadn't worked so hard to minimize engine vibration, so it gave a shaky ride. Sometimes parts shook themselves loose; sometimes they broke violently as they succumbed to metal fatigue.

Arthur Young commented later that the engineers of the Model 42 had tried to design a helicopter as if it was an airplane. The result was a disaster. The "helicopter of the future" soon became the helicopter of the past. The company cut its losses and gave up on it.

As for his own design, it had been conceived from the start to be all-helicopter and became the first helicopter to be licensed for civil use. It was released as the Bell Model 47 and quickly became a huge success for the company. Its big bubble canopy and framework tail became a familiar sight over cities and farms. Civilian versions did everything from crop spraying to traffic reports.

The armed forces bought large numbers of the helicopter—the military version was known as the H-13 Sioux—and it was often used to evacuate injured soldiers. (It is featured in the opening credits of the TV series *M*A*S*H.*) The helicopter continued to be manufactured until 1974. By that time, around 5,600 had been built.

It sounds like a success story for Arthur Young, but he didn't see it that way. He was too much of an artist and dreamer to be excited about the technical accomplishments. By the time his Model 47 was finished, he had a strange epiphany: all his efforts had been misplaced, and his recurring fantasy about floating through the sky had never been about flying an aircraft, but about flying his soul through the universe.

At the height of his engineering fame, Arthur Young abruptly quit Bell Aircraft and went off to teach cosmic awareness, educating other dreamers about the "winged soul" or "psychopter." He moved to California and set up the Institute for the Study of Human Consciousness, incorporating science, world religions, astrology, and a wide range of what we'd now consider New Age ideas. It was certainly a less risky pursuit. Whether Jupiter and Mars influence a person's life or not, they are a lot less likely to fall out of the sky and end it.

Both Arthur Young and his co-workers agreed that he made a big mistake, although they didn't agree what the mistake was. Young thought it was creating the helicopter; his colleagues said it was abandoning it.

According to some, the moral of the story is "Follow your dreams, wherever they take you." Others might say, "Don't give up your day job."

FLYING WITHOUT AN AIRCRAFT

Flying in a helicopter or plane is all very well, but how do you float freely in the air, with no aircraft, no balloon, no bulky machinery around you? One Canadian inventor figured that out.

Jean St-Germain was an inventor from Drummondville, Quebec. We met him in the chapter on food and drink: his first successful invention was a baby bottle that used disposable bags to store the milk. But the baby bottle business was only a passing phase for St-Germain. He was much more interested in flying.

Like Arthur Young, designer of the Bell Model 47 helicopter, St-Germain was fascinated by the idea of tiny, personal aircraft. He didn't want a vehicle that insulated the pilot from the air all around, but one that gave a powerful sensation of flight. St-Germain made himself an expert in skydiving, parachutes, and hang gliders.

In the 1970s, he designed and sold one of the earliest ultralight planes—the single-seat "Moto-Plane." It looked only half-finished, with a sheet metal nose and an open-framed tail.

His next ultralight was even more stripped-down, resembling a go-kart attached to a model glider. He named it the "Raz-Mut," based on the French phrase "rase-mottes" meaning mound-clipper. (The English equivalent would be hedgehopper.)

St-Germain took the Raz-Mut for an unregistered, illegal test flight or two. A government inspector was appalled by the minimalist plane and declared it unfit to fly. St-Germain disagreed, but his attempts to fight the bureaucracy were unsuccessful. Unable to fly the plane legally himself, he presented his micro-plane to American aviation fans in 1977, at the Experimental Aircraft Association Annual Convention and Fly-In, in Oshkosh, Wisconsin. It proved a minor hit and received press coverage. He skirted the regulations by offering the plane in kit form. A handful were purchased and built.

The small airplanes didn't satisfy St-Germain's appetite for flight. He spent a while playing with gyroplanes, a type of aircraft that had been popular since the 1920s. It resembles a helicopter, but its rotor spins freely like a sycamore seed, and it flies more like a small plane. It still wasn't the aircraft St-Germain was looking for. He wanted something that would let him fly straight up into the sky and float there, with the minimum of machinery around him—some kind of helicopter.

In the early 1970s, the inventor built his own prototype helicopter and took it for a few short test flights. That experiment led to more legal trouble. Not only was his helicopter unregistered, but St-Germain didn't have a helicopter licence. Once again, government officials ordered him to stop his flights.

St-Germain refused to give up on his ambitions. He continued to focus on the experimental, ultralight market, and designed a new helicopter that seemed shockingly minimalist. Named the "Hélipac," the vehicle looked like an office chair hanging below a helicopter rotor. A single control stick hung in front of the pilot, but nothing else stood between the pilot and the sky—the chair didn't even have arms.

In 1978, St-Germain returned to the Oshkosh experimental aircraft event and showed off his latest flying machine, taking the micro-helicopter up for a few short hops. It's said the device led to deals with Japanese and British companies and investigations into the possibility of a true "backpack helicopter" that could be worn like a jetpack, but it's not clear what happened to the idea after that. While a helicopter might seem at first to be a liberating alternative to the car, the reality is a little different. Helicopters generally need much more maintenance than a fixed-wing plane to continue working reliably. For some helicopters, each hour of helicopter flight time can matched by twenty hours of maintenance. Perhaps this was part of the reason St-Germain kept looking for other ways to fly.

Many of the inventors who were bitten by the flying bug seem to have had the same dream—to fly in the air with nothing between the flyer and the ground. Despite some impressive efforts, nobody quite managed it—until St-Germain's creative mind came up with an unconventional solution—the "Aerodium."

Every powered aircraft has the same problem. To lift a person into the air, you need machinery. That means you also need to lift the machinery into the air.

St-Germain looked at the problem in a new way. If you want to fly, but you don't want to have a lot of machinery around you, leave the machinery on the ground. A powerful fan can blow people into the air—and the technology is easy. It's just a vertical wind tunnel. And that was the Aerodium.

Of course, a setup like this won't allow people to fly off over distant hills, but it does let them experience the sensation of flying in the air like a bird.

His invention proved hugely popular and provided a big infusion of cash for St-Germain, whose eccentric life had been a series of financial ups and downs. Today's "indoor skydiving" centres are based on St-Germain's invention.

St-Germain's ambitions had strange parallels to those of helicopter designer Arthur Young. Not only did both men pursue the dream of free flight, but they also embraced spiritual ideas. One of St-Germain's pet projects was a huge stone pyramid that was supposed to focus cosmic powers to recharge a person's energy. A nearby "transitional ring" then boosted emotions, and a geomagnetic cylinder improved physical health. He also built a giant crucifix, with 2,000 lights—perhaps trying to cover more spiritual bases.

St-Germain considered himself better at inventing than running a business. When he got money, he usually burned through it. At one point, he had a large tax bill, which he said he was unable to pay. Officials got a court order to open his safety deposit box. Inside, they found fifty pennies. When the confused tax officials asked St-Germain what the coins were for, he told them it was his emergency fund. With that money, he could buy a pencil and a notepad—the only tools an inventor really needs.

CHAPTER 24

WHAT'S THE BIG IDEA?

SOME ideas are very specific—like figuring out the right material for a light bulb filament or a new design for a paper coffee cup.

But some new ideas are much grander in scope. These are the inventions that revolutionize society, changing the way we all work, and making everything better, easier, and more efficient.

Given the resistance of most people to change of any kind, the visionaries who want to make these big ideas happen usually face a very rough ride.

A TWENTY-FOUR-HOUR JOB

You might wonder where inventors find time to come up with their ideas—after all, there are only twenty-four hours in the day. But having twenty-four hours in the day was itself an invention—and when the idea was first presented, nobody liked it.

Back in ancient Egypt, they used sundials to tell the time. The oldest one found so far is about 3,500 years old, and it has a scale with twelve even markings for the hours of daylight. Some unknown inventor a very long time ago decided that twelve would be a good number of daytime hours. Nobody knows why that number was chosen. It might have been because the Egyptians liked counting by twelves—when they counted on their hands, they would use a thumb to count their finger joints—three joints on each finger, making twelve in total. Then again, it might have been because there are twelve months (lunar cycles) in a year, so it seemed intuitive to divide the day up in the same way.

If you're going to divide the day up into twelve hours, it makes sense to do the same with the night, but that's more difficult to figure out. The Egyptians did it by watching the movements of various key stars. Once they had worked out a system, they had a day made up of twelve hours, followed by a night made up of twelve hours.

But their hours were a little different from the hours we use today.

One problem with dividing up the day using sunlight is that the day and night aren't the same length. In summer, daytime is longer and nighttime is shorter, and in winter it's the other way around. If you always have twelve hours of daylight, the hours get longer or shorter at different times of the year. In summer, a daytime hour might be seventy-five minutes long, while a nighttime hour would only be forty-five minutes. And the further north you go, the worse the problem gets. During an Arctic summer, there would be no nighttime hours, and a daytime hour would be 120 minutes long.

An ancient Greek astronomer named Hipparchus had a great solution. Why not combine day and night into one big block, and divide it into twenty-four hours that stay the same length all year long? That way, if you wanted to see how many hours a job took, you wouldn't have to measure it in "summer hours" or "winter hours" or "early spring hours." You could just say "hours" and everyone would understand.

The reaction from the public was not great. These geeky "fixed hours" sounded much too complicated. No, they would rather stick with the system they already knew.

And they did stick to it—until the Middle Ages, people were still measuring time by the sun, using the same elastic hours the Egyptians used. It made some things simple. The sun always rose at the start of the first hour, reached its highest point in the sky at the sixth hour, and set at the twelfth hour, winter and summer. But hours weren't much use for measuring anything else—the length of an hour was never the same from one day to the next, or from day to night.

To measure smaller amounts of time, some sundials had their hour marks subdivided into forty "moments"—which was pretty much the smallest movement you could see on a typical sundial. But that didn't help, because the "moment" was as stretchy as the hour—in summer, it might be ninety seconds, but in winter it was only forty.

Fixed-length hours didn't become popular until mechanical clocks started becoming a regular fixture on church towers. The hours on a clock wouldn't stay in sync with the sunrise and sunset, so people finally threw in the towel and used the system Hipparchus had suggested. From then on, the length of an hour stayed the same and the times for sunrise and sunset moved around. The official start for each date could not begin at sunrise (because it kept changing), so it was arbitrarily set at midnight—the new "day" began in the middle of the night.

There have been many inventions that didn't become successful until after their inventors had died, but Hipparchus's invention of fixed hours must take the record for the slowest adoption—more than 1,400 years from its invention to the start of its acceptance in the Middle Ages.

Sorry, Hipparchus. It was a good idea. It just wasn't the right time.

FROG BATTERY

Important inventions aren't always the result of cool-headed research. Sometimes inventors just want to prove the doubters wrong.

Luigi Galvani was an Italian scientist who became famous in the late 1700s for his amazing animal experiments. He could take the legs from a dead frog and, by poking at them in the right way, make them twitch.

Galvani had a big idea about his experiments. He believed he had discovered a new kind of electricity, a vital power stored in the fluids of the frog's body. He was making dead tissue move like live tissue, and he viewed this animal electricity as something unique. He believed he had found a deeply mysterious and special power—a sort of life force.

Another Italian scientist, Alessandro Volta, was more skeptical. He suggested that Galvani might have things backwards. He'd noticed that Galvani's probes were made from two different metals, iron and brass. Perhaps the mixture of metals was somehow creating electricity, and that was making the muscles twitch. Volta had tested this himself—when he tried prodding at frog legs using two different metals, the legs twitched, but when both probes were the same metal, nothing happened. It shouldn't work that way if the electricity was already inside the frog.

Galvani wasn't impressed. He didn't think Volta knew what he was talking about.

Volta was so irritated by Galvani's reaction that he set out to prove his point. He would make frogless electricity!

Volta built a big stack of metal disks, using two metals that were very different, zinc and copper. He put them in layers, copper, then zinc, then copper, then zinc again, with a disk of brine-soaked cardboard between each disk. He stacked them up as high as he could manage without the pile falling over.

When Volta attached wires to each end, and touched it to wet fingers or his tongue, he could feel the electricity, and with not a frog in sight. There was no need to talk about "animal electricity"—there was only electricity. Take that, Galvani!

In proving his point, Volta had invented an important new technology: the big pile of disks was the first true battery. That's why, in French, a battery is still called a "pile." It's also why today, all over the world, electrical charge is measured in volts, not galvanis—although Galvani's name appears in "galvanized" metal—a metal-coating applied using a battery.

Was Galvani persuaded? Not at all. He was still convinced that animal electricity was a real and important phenomenon. He had many followers who backed him up—they liked the idea that living things were powered by a special life energy—although they were also intrigued by Volta's approach of making more power by using a bigger battery.

This mix of ideas led to a series of bizarre experiments, where Galvani fans would stack a series of frog legs together to make a high-powered "frog battery." Sometimes researchers wired carefully sliced portions of frog muscle together; other times they used the whole frog. It was an amphibian version of *The Matrix*: the frog had been turned into a battery.

Other scientists, mostly Italians, tried other variations, using different animals and different body parts, in a series of gruesome experiments. Giovanni Aldini slaughtered an ox and used its head to create electricity. He then moved up to three ox heads wired in series. Another scientist, Carlo Manteucci, made frog-muscle batteries, then followed it up by wiring together dead eels, pigeons, and rabbits. He discovered you could even use living pigeons to make electricity, provided you made small holes in them first.

Again, it worked, but only because of the mix of metals used. The animals' only meaningful contribution was the salt water in their bodies.

Writer Mary Shelley was fascinated by Galvani's work and by the prospect of using this mysterious animal electricity to reanimate dead tissue. It was an influence on her novel *Frankenstein*. There was no lightning storm in her original book—it was all done with "galvanism."

In the real world, animal electricity proved to be a dead end (in more ways than one), and all the progress in electricity came from Volta's battery invention, which itself had come from the scientist's desire to win an argument.

As the frog might have said, "It's not my volt."

PITMAN SHORTHAND

Isaac Pitman was one of those people who didn't like English spelling and was always saying, "Why can't they just spell it the way it sounds?" However, he did more about it than most.

In 1837, when Pitman was twenty-four or so, he invented a new system of shorthand. Shorthand wasn't new—various systems had been used since ancient times. But Pitman's system was unusual. Most earlier shorthand systems provided a list of abbreviations for common words and phrases, and methods of shortening others. ("If u cn rd ths u cn gt a gd jb w hi pa!") But Pitman's system was unusual—it was designed to capture all the common sounds in English, using a set of curves and lines that could each be written with a single rapid stroke. One curve stood for the sound of "sh" and another stood for "th"—in fact, two curves stood for "th": one for the voiced "th" in words like "this" and the other for the unvoiced "th" in "thing."

Pitman was working as a schoolteacher at the time. He printed his system in a small pamphlet and taught it to his students. They had no difficulty learning it, and they had fun using it to write secret letters back and forth.

Pitman knew his shorthand would be good for court reporters and secretaries to transcribe speech, but his sights were set higher—he wanted his style of writing to replace the old alphabet. With his "phonographic" writing, every word would be spelled exactly the way it sounded.

Pitman promoted his system well. He printed publications and set up a correspondence course—the world's first—to teach the system to others. It caught on fast. Court reporters loved it—with practice, they could easily write at the speed people spoke. Of course, it was ideal for taking dictation, but it was also sometimes used for business correspondence—letters written in shorthand had an added layer of privacy and also used less paper than ones written out in full, and that saved money on postage.

Today, shorthand is remembered as a skill practised by 1960s office secretaries, but in Pitman's day, the users were mostly competitive young men. Shorthand organizations held speed tests to see who was the fastest.

Language reformers bought into Pitman's dream to make his shorthand system the primary way of writing English. And they didn't want to stop with English. Variations in the system were designed to write

French, Russian, even Latin. They foresaw a time when people would be able to write to each other all around the world using a single, logical phonetic language. They might not know what a person in another country was writing, but they would know exactly how to read it aloud.

Pitman republished popular books, entirely printed in shorthand. Because the writing system was so compact, a large volume often ended up small enough to slip into a pocket. He printed shorthand magazines, too. The movement was gathering some momentum among the "geeks" and early adopters of the day. It started to seem possible that Pitman's writing system might actually become a standard writing system for the world. But things started to go wrong.

As an Englishman, Isaac Pitman had focused most of his attention on the British Empire and hadn't worried too much about protecting his invention in the United States. That was a mistake. Similar shorthand systems started popping up across the US. Most of the American systems were blatant copies of Pitman's system—usually the same system with a different name attached. Some claimed to be faster, easier, or "more scientific" than Pitman's original.

Isaac Pitman sent his brother Benn to the United States to assert his rights there. But by then, another system, Gregg Shorthand, had also become popular in the United States—it was fine as a shorthand system, but it didn't capture the full range of sounds that Pitman did. Amid the confusion of different systems, Pitman missed the chance to make his own shorthand system the dominant one in the US.

Isaac Pitman's restless personality also got in the way. After inventing a good system, he couldn't resist making tweaks and changes, and one of his changes involved changing the order of all the vowels (represented as dots and dashes in different positions). It was rather like being told that, from now on, the letter "e" means "a" and vice versa. Thousands of people had taken the time to learn his system, and some were busy teaching it to students. They saw no good reason for the change and resented having to adjust to a new system on Isaac's whim.

Shorthand also led to a more personal type of problem for Pitman. His organization had always encouraged a network of pen pals. Pitman's wife was suspicious about one stack of letters her husband had accumulated. She couldn't read shorthand herself, so she asked one of Pitman's students to read the letters. It turned out they were from a woman named Martha Watts, who had been sending Isaac Pitman love letters using the secret writing system.

But according to famous Pitman writer George Bernard Shaw, the biggest problem may have been shorthand's emphasis on speed over simplicity. As the most enthusiastic and competitive writers developed new tricks and techniques for writing Pitman's shorthand, the system became increasingly fast, but it also became more complex and harder to learn.

Pitman shorthand had originally been printed on a pamphlet and taught to a class of small boys in a few days. Once you learned its forty-odd symbols, you could read anything. But in a few decades, it had morphed into a system that needed to be taught with full-sized textbooks, containing countless rules and exceptions to rules, all to reduce the number of written characters and increase speed. Instead of taking days to master the system, it took months. It was an excellent system for recording speech—shorthand written by pen is still much faster than typing—but it was no longer a good candidate for a new universal writing system.

Later, as more women learned shorthand for taking dictation in the office, the men who had been so enthusiastic in the early years started to drift away. By the sexist standards of the day, both shorthand and typing started to be viewed as "women's work," and therefore beneath the dignity of a man.

Pitman's shorthand invention remained popular for decades as a valuable office skill, but he missed the chance to make it a new way of writing the English language.

HELLO, GIRLS

If you invent a new technology, you might have to invent behaviours to go with it.

Take the telephone, for example. Once it had been invented, tested, and refined, the inventors realized they needed to give people precise instructions on how to use it politely. If you're a caller who's just been connected to your call-ee, what are you supposed to say to start the conversation? Alexander Graham Bell and his telephone rival Thomas Edison gave this question serious thought.

The phones were noisy, and it was hard to hear voices, so if one person called another on the telephone, it often involved a lot of literal calling. Communication was like shouting from one ship to another.

Inspired by the similarity of hailing a ship, Alexander Graham Bell suggested that a good phone greeting would be "Ahoy-hoy!" That didn't catch on, although Montgomery Burns used it in *The Simpsons*—a sign of his advanced age.

Edison was running a competing phone system. His telephone guidebooks told people to begin the call with "Hello!" That advice was also given to his telephone operators.

Hello might not seem a very inspired choice, but in those days, hello wasn't used the same way we use it today. It was something you said to get someone's attention across a valley or when calling from the street to a fourth-floor window. And if you were surprised, you might exclaim, "Hello! What's this?" But in the late 1800s, it was never a daily face-to-face greeting. Saying hello to someone on the street would have been like walking up to them and greeting them with a giant wave of your arms.

That changed overnight. Hello was suddenly the greeting used by hundreds of telephone operators. These were generally young women chosen for their pleasant speaking voices. Telephone operators quickly became known as "hello girls." The word "hello" felt as fresh and charming as the voices saying it. Before long, hello had taken off as a popular, fashionable greeting between people, replacing fusty old standards like "good day."

Talking to a woman required an introduction in those days, so some men were enchanted at the novelty of picking up one of these newfangled telephones and hearing a lovely female voice at the other end. This slightly creepy fascination even inspired a song about a man who falls in love with a hello girl. The title was "Hello! Ma Baby," and it became a huge hit. The song features prominently in the famous 1955 cartoon about the singing frog, *One Froggy Evening*, but it dates from the end of the previous century and was first recorded on a wax cylinder. Both the cylinder and the word "hello" as a casual greeting were invented by Edison.

CHAPTER 25

CREAM OF THE CRACKPOTS

OVER the years, there have been countless determined but delusional inventors proposing loopy ideas that make no sense or wouldn't work.

But there are a few inventors in the crackpot category whose work outshines the rest. These are the rare individuals whose inventive interests are eccentric, eclectic, and enthusiastic. They are impractical but prolific, producing a torrent of hopeless ideas.

There isn't space to look at them all, so we'll present just two shining examples.

Say hello to the cream of the crackpots.

THE WORLD'S WORST INVENTOR?

Einstein had worked as a patent office official. He went on to discover relativity, and the physics that would lead to the atomic bomb.

Arthur Paul Pedrick also worked as a patent office official. He became one of the most prolific and eccentric inventors of all time.

Pedrick had strong opinions about Einstein's work at the patent office. He thought many of the world's problems came from Einstein doodling his dangerous equations, "instead of getting on with the work he was being paid to do."

In truth, Pedrick hadn't liked working in a patent office any more than Einstein had. He had been hired in 1947 as a patent examiner in Britain, looking at patent applications and researching to see if similar ideas had already been patented. He found the job exhausting and had a hard time making decisions. Since making decisions was what the job was all about, he was eventually dismissed in 1961 for "inefficiency." He would have been around forty-three.

After leaving his job, Pedrick moved in with his aunt, who had a good-sized house near the seaside on the south coast of England. His mother and another aunt also lived there, and they were soon joined by Pedrick's cat and feline muse, Ginger. From this creative headquarters, Pedrick started to write patent applications of his own, sending a steady stream to the patent office.

Pedrick watched TV and read the newspapers, and whenever he saw a problem with the world, he'd dash off an invention to solve it and mail it to the patent office. As a former patent examiner, he knew how to write his applications in a way that would get them accepted. His first patents were quite technical, involving improvements to engines and machines, but as time went on, he started to hit his stride. His applications became increasingly eccentric and often included long commentaries with many personal details. Some of his patents described ways to create nuclear fusion, although it wasn't an area where he had any formal education or work experience. In one of these patents, he included a discussion about how his cat preferred corned beef to any of the tinned cat foods. The rules of the patent office meant that all these meandering additions had to be included in the final patent.

Pedrick's inventions covered a very wide range. Some dealt with minor domestic problems. He offered an improvement to the tea strainer, with three sets of fine wires arranged in triangles, rather than the usual square grid. He seemed to think the triangles would do a better job of filtering out tea leaves.

For those who wanted to hang a picture but had trouble getting started when hammering a nail into a wall, he invented a nail with a small explosive charge inside it attached to a needle running inside the nail. The first tap would set off the charge, driving the needle into the wall and holding the nail in place. The nail could then be pounded the rest of the way into the wall using old-fashioned hammering. When all you need to do is make a starter hole, this feels like the nuclear option.

Other inventions were global in scope. He patented a pipeline to transport Antarctic ice (packed into snowballs) to the deserts of Australia for irrigation. He also patented floating farms to feed the world's growing population.

He filed a total of 162 patents in the UK, which cost him thousands of pounds to submit. None became commercial products. But for Pedrick, it wasn't just about the invention—the patents themselves seemed to be a way for him to share his experiences and his views about fixing the world.

REINVENTING GOLF

Arthur Pedrick liked golf and filed many golf-related patents—although the ideas he came up with suggest that he might not have been a very good player.

Many of his golf inventions address the problem of a player hitting the ball only to see it spinning off at the wrong angle. One patent was for a golf ball with tiny flaps, attached with magnets. When the golfer's bad stroke caused the ball to spin, the centrifugal force would cause the flaps to fly open, correcting the problem. The ball also contained a device to reflect radio waves, so the golfer, equipped with a compatible radio unit, could find a ball that was lost in long grass or bushes.

Pedrick seemed to view golf more as a set of technical problems than a sport. Other golf inventions involved the tee—the little spike that supports a golf ball when the golfer hits it. Pedrick's idea was to create a tee connected to a set of lights and photocells. It would also be connected to a supply of compressed air that had its own miniature photocell. When the golfer took a swing at the ball, the sensors would detect the accuracy of the stroke and make a judgment that seems like it would be at the cutting edge of modern AI: if it looked as if the club would hit the ball badly, the system discharged compressed air through the tee, sending the golf ball flying briefly upwards, so the golf club would miss it entirely. That way, the player would be saved the effort of having to look for a lost ball.

In 1974, he decided that the real problem with constantly slicing or hooking the golf ball came from the fact that the ball was resting on that annoying tee. His solution (on paper only) was to attach a Van de Graaff generator, levitating the ball by electrostatic forces, so it could be hit with greater ease.

During a different week, Pedrick came to the conclusion that the problem wasn't the tee or the ball, but the club. He patented a new design, which contained a number of vertical rubber blades. The blades would extend when the golfer took a swing, "resisting the tendency of the ball to roll across the face of the club." Because the problem was the ball and the club, not his terrible golfing skills.

PEDRICK'S TRAVEL PLANS

Arthur Pedrick followed the news carefully, and many of his inventions seemed to be inspired by news reports or films. Here are just a few of the inventions he patented that seemed to be inspired by current events.

A HAY-POWERED CAR

He thought of this during the oil crisis of the 1970s. The vehicle was basically a car with a horse strapped behind it to provide the power, similar to another invention we've seen, P. A. Barnes's horse-powered minibus. In Pedrick's version, turning the key in the ignition was supposed to give the horse's rear an electric shock, encouraging it to move. The driver could then steer the car normally. Applying the brake pedal tightened a rope that tugged at the horse's halter. A small trailer behind the car was filled with food for the horse.

AIRSHIP MOORING TOWER AND MEDITATION ROOM

This invention, from 1968, seems to have been inspired by the Beatles' trip to India. Pedrick envisioned a growing class of jet-setting hippies who wanted to travel the world by airship and never wanted to touch the ground. He patented a tower with an airship mooring mast, a hydraulic elevator, a revolving restaurant, and a clear globe at the top of the tower where a passenger could sit and engage in transcendental meditation. Sounds groovy.

TRIPLE-SIZED JUMBO JET

Back in the real world—although not far back—Pedrick patented an aircraft. In 1970, the TV and newspapers were full of images of the new Boeing 747 jumbo jet, which was then making its first commercial flights. Pedrick patented his own even larger aircraft, consisting of two jumbo jets joined together, with the nose of a third one stuck in the middle like an aerial trimaran.

Of course, this aircraft was never built, but surprisingly the patent did produce results: a few years later, Airbus was working on its own ultra-large airliner. Their competitor Boeing tried to patent an even bigger airliner made from three planes stuck together. Boeing's patent was unsuccessful because Pedrick and his cat, Ginger, had already patented a similar idea. It's a rare case where one of Pedrick's patents actually had a real-world application—even if it was only foiling another patent.

EVEN BIGGER, CRASH-PROOF JUMBO JET

In 1974, he thought about making a giant aircraft safer. He might have been inspired by recent hijackings of 747s, or perhaps by the upcoming film *Airport 1975*. Pedrick proposed a novel alternative to the jumbo jet. Again, it involved strapping a group of jumbo jets together into a single, very wide aircraft. The new twist was the altitude. It would be flown just a few feet above the ground for maximum safety (of those in the plane). At the first sign of trouble, the plane could drop down a few feet and land, or, if it was over the sea, descend on built-in floats.

Although if you talk to pilots about the safest way to fly, they will not usually suggest staying as close to the ground as possible.

CATASTROPHE AND A NUCLEAR FLAP

Pedrick had a background as an engineer. He served in the navy during World War II. He had one very bad experience when his ship, the HMS *Dido*, was attacked by German dive bombers. He was serving in the Pacific when the first atom bomb was dropped on Japan. These events gave him a strong pacifist outlook. He was constantly thinking about the threat of nuclear war.

Perhaps Pedrick's strangest patent, and one of his best known, was for a cat flap. In his patent, he explained that his cat, Ginger, entered the house through a cat flap, but his neighbour's black cat, younger and fitter than Ginger, would often dash through the cat flap ahead of Ginger and eat Ginger's food.

To exclude the neighbour's pet, Pedrick proposed a device that could monitor coloured light coming from below the cat flap. The flap would remain locked if the colour was black, like the fur of his neighbour's cat, but if the reflected colour was ginger, the system would activate, unlocking the cat flap.

But what sounds like an almost-practical domestic invention quickly turns in an entirely different direction.

In the same patent, the inventor proposed putting the same colour-detecting sensor on a series of orbiting satellites carrying nuclear weapons. If the satellites spotted a nuclear explosion (presumably by the distinctive hues of the blast), they would automatically fire their weapons at the aggressor. Once this system was in place, world peace would be guaranteed, and the nations of the earth could decommission their weapons and use the nuclear fuel for peaceful purposes.

His patent had the catchy name "Photon Push-Pull Radiation Detector for Use on Chromatically Selective Cat Flap Control and 1000 Megaton Earth-Orbital Peace-Keeping Bomb."

PEDRICK'S SAFER WORLD

When disaster stories were in the news, Arthur Pedrick often responded with an invention to keep people safe. In 1974, a yacht named *Morning Cloud,* belonging to former British prime minister Edward Heath, was caught in a storm. The yacht sank and two of the crew members drowned. Pedrick responded by inventing a rubber ring that could inflate around a yacht and give it extra buoyancy in an emergency.

Fictional dangers also inspired Pedrick. After the film *The Towering Inferno* came to the cinemas at the end of 1974, Pedrick thought about how to extinguish a fire in a burning skyscraper. His response was a patent for a fire curtain that could be installed around a high-rise building. If a fire were to break out, the curtain would drop around the building, enclosing it completely and preventing air from entering. Without oxygen, a fire can't burn.

But what about the people? Wouldn't they be asphyxiated? Pedrick considered that, too. The curtain has holes, allowing air to enter certain key rooms. The residents can flee to those rooms and be safe. It's not clear what happens if it's one of those rooms that's on fire.

Sadly, this was Pedrick's final patent. His beloved cat, Ginger, died of old age in 1976, and Pedrick followed soon after. The staff at the patent office missed him. Those who worked there used to look forward to his latest eccentric offerings.

CLARA LOUISA WELLS

Although the most eccentric patents seem to come from male minds, there have been some women who gave those men a run for their money.

Clara Louisa Wells was a crackpot inventor superstar in the early twentieth century. She was a successful American writer who lived and wrote in Europe, and like Arthur Pedrick, was interested in the peace movement.

Wells was an amateur inventor, and her patents combine grandiose ambition with an impressive lack of practicality or technical knowledge.

BOAT SHOE LIFE JACKET

Safety was often on the mind of Clara Louisa Wells, and like Arthur Pedrick, she seems to have been inspired by the latest news stories. In May 1915, the ocean liner *Lusitania* was sunk by a German U-boat. A few months later, Wells had patented an invention to save passengers facing such dangers.

Her invention was an elaborate life jacket—although perhaps life-ensemble would be a better description. When a passenger of a ship suspects that their vessel is about to sink, they slip on a number of matching accessories. A custom air-filled hat will keep the head buoyant—it looks exactly like a small bowler hat. The passenger's body is adorned with a mid-length coat: a set of small cork floats are placed around the hem, and more corks encircle the collar. The passenger's feet slip into matching cork shoes, which are attached to two small boat-like skis.

Her illustration shows a dapper passenger standing calmly on the water, holding two balancing poles, with a cork float at each end. Since the boat shoes allow the person to stand on the water, the hat and coat seem unnecessary.

Wells cared as much for dogs as for people, and her patent includes a dog version of the design, with a dog-sized jacket, four boat shoes instead of two, and a balancing pole Rover can carry in his mouth. The only thing missing is a training manual so the dog can learn how to use it.

An electric motor is shown strapped to the passenger's back, although it's not explained where the electricity comes from or what the motor connects to—perhaps it is some kind of outboard motor, in case the wearer decides to take off their floating footwear and swim instead.

Overall, it's a complex solution, and those facing the heaving waves of the North Atlantic might feel safer sitting in an inflatable raft than trying to stand upright on tiny floating shoes.

Wells points out that her water suit is not just for survival after a shipwreck. It can be used in other situations. She suggests that it would make it easy for people to carry messages across rivers—she wisely advises that the messages first be sealed in waterproof envelopes.

The dog suit also had other applications: she believed it would provide a good way to deliver food to flood victims, "since the dogs would not fail to try to reach those they love." She ignores the fact that being able to send the dog to a victim presumes the animal has already abandoned those they love.

GAS GUZZLERS

During World War I, troops were constantly at risk of being gassed. In 1915, Clara Louisa Wells considered the problem and patented her solution, which involved sucking up all the poisonous gas—a vacuum cleaner for the air.

Her system was to be mounted on two aircraft, which would fly through the poisonous clouds. The front aircraft was equipped with an air-gathering device, which resembled three huge trumpets projecting from the nose of the plane. A long rubber hose stretched from the first plane to a second plane. It carried a giant tank atop its fuselage, which would store the poisonous gases.

It is not explained how the aircraft is able to suck up only gas and not the rest of the surrounding air—a rather important detail.

Once the storage tank is full, the plane releases the gas "far away" and the aircraft returns to the battlefield to suck up the next batch of gas.

The details of the planes are rather vague, but she does offer some suggestions to the engineers who will build them. "The airplanes used for this purpose must be very powerful and specially designed. They will need large wings and strong engines. The pilots must be equipped with respiratory masks, which will protect them against asphyxiating gases."

She was certainly right about her aircraft needing strong engines. Wells's aircraft appear so aerodynamically challenged that they would have had difficulty staying upright on the tarmac, never mind getting into the air while towing a length of heavy hose between them.

However, her system didn't have to use aircraft. In the same French patent, she offers a second variation using two trucks—the front one was equipped with the same strange air-sucking horns. A long hose was connected to a second truck carrying a tank to hold the poisonous gas. It's not clear why the hose arrangement is necessary. Why not put the gas tank on the same vehicle that carries the horns? It would have worked just as well—which is to say, not at all.

EXPLORING MADE EASY

As a world traveller herself, Clara Louisa Wells was interested in the welfare of her fellow explorers. In the early 1900s, newspapers and magazines were filled with the exploits of explorers going into dangerous parts of the world, from the wastes of Antarctica to the jungles of Africa.

She worried about the safety of the brave explorers, "who endeavour to reach these regions, in frail vessels, steamers, or solitary balloons, hazarding all with little probability of success." These thoughts led to her wildest and craziest patent.

She proposed a travel network between the most dangerous sites on earth, both the very hottest and the very coldest, using cable cars that would hang below cables, with tethered balloons floating above them. The network would cross oceans. The cables would be strung from the highest points on islands or from towers emerging from the sea.

She wanted to improve the temperature in the world's most inhospitable places, so her network also included a series of pneumatic tubes, pumping cold air to hot places and vice versa. (She imagined the air would somehow stay warm or cool even after travelling a quarter of the way around the planet.)

She also had a second method for regulating heat at each location. She knew (or at least, believed) that the north and south poles also were the sites of volcanoes and hot springs, which could be tapped for heat. She also knew that the hottest parts of the world, around the equator, had many snow-capped mountains, which could be used for cooling.

In her scheme, visitors to each station could be kept comfortable by changing their elevation. In the blazing heat of Africa, a travel centre would be placed on a snow-capped mountain. If people were too hot, they could cool off by ascending to the peak on a sort of elevator. At stations in polar regions, a person who was cold could get warm by going up or down into the hot springs and volcanoes conveniently situated nearby. Vents could also pump warm or cold air from one altitude to another.

Once the work of building this network of cables and tubes was complete, the most dangerous parts of the world would be transformed into giant holiday resorts. Each centre, she said, will "be furnished with convenient means of access; with different tramways; with aerial locomotion; with railways, also suspended vehicles and the like; as well as with hotels and restaurants, when there are none in the vicinity."

And there was no need to worry about any risk from those active volcanoes in the polar regions. Each one would be surrounded by a network of "defences, ditches, and canals" to divert any rivers of lava.

Presumably, once the network was built, the "courageous men" who had formerly tried to reach these remote areas in their flimsy boats and balloons could get there safely in a few hours by cable car, enjoying dinner and a comfortable room at the other end—although this perhaps misses the point of being an explorer in the first place.

More importantly, if these areas are so dangerous that the bravest explorers can't reach them, then building this travel network would require impossible, superhuman powers.

Wells often focuses on easy problems, like how to attach a balloon to a cable, and waves away the biggest ones, like how to control the lava from an erupting volcano or how to build vast towers in the middle of the ocean.

Like all the best crackpot ideas, her invention is wild, grandiose, and utterly impractical.

CHAPTER 26

FINALLY . . .

WALK through a living room, or a mall, or an office and you will see many objects made by your fellow humans. Every one of them started out as someone's new idea. Some objects, like computers or mobile phones, draw on hundreds of patents.

And the ones you see are just the success stories. For every invention that was embraced by the public, there are thousands that remain invisible, ideas that never quite made it.

It's entertaining to laugh at the failures, and—in hindsight—easy to judge the good from the bad. But as we've seen, some ideas that looked ridiculous to the inventor's colleagues became important breakthroughs, while more than a few great-looking ideas were sunk by twists of fate and now appear strange or comical.

The progress we enjoy comes about by trial and error. Without all the errors, we would never see the few brilliant successes.

But when the odds of a new idea making money are so low, why do the inventors keep at it? Perhaps it's for the same reason that unpublished novelists keep writing and amateur painters keep painting. They want to express themselves, to make a difference by putting something new into the world. They know they are taking a chance, and they might lose money or be mocked for their efforts, but they are courageous enough to try.

Whether it's Bette Nesmith with her Liquid Paper or Arthur Pedrick with his peacekeeping cat flap and radiation detector, these creative individuals aren't just craftspeople. They are problem solvers, disruptors, social critics, and artists.

So, if you've got a great idea, believe in it. Do what the other inventors have done: get out there and promote it! Pour your effort into it! Get a patent! Spend your savings! Quit your job! Mortgage your house!

What could possibly go wrong?

ACKNOWLEDGEMENTS

My special thanks to Lachina McKenzie and Catherine McKenzie for their invaluable feedback and help in proofreading this book.

Thanks also to Alan Eliasen for his help with the parachute calculations, to Professor R. John Williams at Yale for historical information on the Printator device, and to Kevin Ptak for information about his faster-than-light bicycle.

BIBLIOGRAPHY

Anderson, Clive. *Patent Nonsense*. London: Michael Joseph, 1994.

Brown, G. I. *The Guinness History of Inventions*. Bath: Guinness Publishing, 1996.

Chaline, Eric. *History's Worst Inventions and the People Who Made Them*. London: New Holland Publishers, 2009.

Hart-Davis, Adam. *Eurekaaargh!* London: Michael O'Mara Books, 1999.

Morris, Neil. *From Fail to Win: Learning from Bad Ideas*. Chicago: Raintree, 2011.

Murphy, Jim. *Weird and Wacky Inventions*. New York: Crown Publishers, 1978.

Murphy, Jim. *Guess Again: More Weird & Wacky Inventions*. New York: Bradbury Press, 1986.